JOURNEYMAN ELECTRICIAN EXAM PREP

2 in 1 Theory and Practice: Ultimate Guide with 500+ Practice Questions and Full Test Simulation for First-Attempt Success

Thomas Farris

Copyright © 2024 by Thomas Farris

All rights reserved. No part of this publication may be reproduced, distributed, or transmitted in any form or by any means, including photocopying, recording, or other electronic or mechanical methods, without the prior written permission of the publisher, except in the case of brief quotations embodied in critical reviews and certain other noncommercial uses permitted by copyright law.

DISCLAIMER

This book is intended for informational and educational purposes only and is specifically designed to assist readers in preparing for the Journeyman Electrician Exam. The author and publisher have made every effort to ensure the accuracy and reliability of the information provided within this publication. However, the information is provided "as is" without warranty of any kind. No responsibility is accepted by the author, publisher, or any persons involved in the creation of this book for any errors, omissions, or inaccuracies contained therein or for any actions taken in reliance thereon. This book does not substitute for professional training or licensed professional advice specific to your local regulations and standards. Always consult a professional electrician and the local electrical code authorities to ensure compliance with all current and applicable codes and regulations.

TABLE OF CONTENTS

PART 1 9
THEORY 9
CHAPTER 1 10
INTRODUCTION 10
1.1 Exam Overview: Purpose, Scope, and Importance 10
1.2 Eligibility Criteria: Educational and Apprenticeship Requirements 11
1.3 Application Process: Steps, Fees, and Deadlines 12
1.4 Exam Format and Question Types 13
1.5 Scoring and Certification Outcomes 13
1.6 State-Specific Regulations and Requirements 14
1.7 Exam Day Essentials: Preparation and Expectations 15
1.8 Study Resources: Books, Courses, and Online Materials 16
CHAPTER 2 17
MASTERING THE NATIONAL ELECTRICAL CODE (NEC) 17
2.1 Historical Context and Objectives of the NEC 17
2.2 Navigational Strategies for the NEC 18
2.3 Key Updates in the Latest NEC Edition 19
2.4 Case Studies: NEC Application in Practical Scenarios 20
CHAPTER 3 21
ELECTRICAL THEORY FUNDAMENTALS 21
3.1 Basics of Electricity: Voltage, Current, and Resistance 21
3.2 Electrical Power and Energy Essentials 22
3.3 Conductors, Insulators, and Semiconductors 23
3.4 Ohm's Law: Principles and Applications 23
3.5 Kirchhoff's Laws in Electrical Circuits 24
3.6 AC/DC Theory: Understanding Current Types 25
3.7 Advanced AC Concepts: Waveforms and Circuit Dynamics 26
3.8 Power Calculations in Various Circuits 27
3.9 Practical Electrical Problem-Solving 28
CHAPTER 4 29
PROFICIENCY IN ELECTRICAL CALCULATIONS 29
4.1 Circuit Analysis: Series and Parallel Configurations 29
4.2 Voltage Drop: Understanding and Calculations 30

4.3 Conductor Sizing and Overcurrent Protection 31

4.4 Comprehensive Load Calculations 32

CHAPTER 5 33

WIRING METHODS AND MATERIALS 33

5.1 Types and Selection of Conductors and Cables 33

5.2 Raceways, Boxes, and Fittings: Installation and Types 34

5.3 Best Practices in Wiring Methods 35

5.4 Comparing Residential and Commercial Wiring Standards 36

CHAPTER 6 37

ELECTRICAL EQUIPMENT AND DEVICE MANAGEMENT 37

6.1 Overview of Key Electrical Equipment 37

6.2 Installation and Use of Switches, Receptacles, and Fixtures 38

6.3 Fundamentals of Motors and Generators 39

6.4 Transformers and Capacitors: Types and Applications 40

6.5 Grounding and Bonding Techniques 40

CHAPTER 7 43

SAFETY AND COMPLIANCE IN ELECTRICAL WORK 43

7.1 Core Safety Principles for Electricians 43

7.2 Personal Protective Equipment (PPE): Selection and Maintenance 44

7.3 Safe Work Practices: Lockout/Tagout and More 45

7.4 Adherence to OSHA Standards 46

7.5 Emergency Response: First Aid and Fire Safety Protocols 46

CHAPTER 8 49

ADVANCED MOTOR CONTROLS AND AUTOMATION 49

8.1 Basics of Motor Controls 49

8.2 Different Types of Motor Starters 50

8.3 Role and Functionality of Relays and Contactors 51

8.4 Introduction to Automation Systems 52

CHAPTER 9 55

DESIGN AND MANAGEMENT OF ELECTRICAL FEEDERS 55

9.1 Introduction to Electrical Feeders 55

9.2 Electrical Feeder Design and Application 56

9.3 Sizing, Selection, and Installation of Feeders 57

9.4 Overcurrent Protection and Voltage Drop Considerations 58

9.5 Grounding, Bonding, and Maintenance of Feeders 59

PART 2 61
PRACTICE 61

Journeyman Electrician Exam PRACTICE (500+ Exam Questions) 61
Specific Questions for each topic 61
National Electrical Code (NEC) Questions 61
Answers 69
Electrical theory 75
Answers 83
Electrical Calculations 89
Answers 96
Expertise in Wiring Methods and Materials 103
Answers 109
Electrical Equipment and Devices 115
Answers 122
Safety and First Aid 128
Answers 137
Motor Controls and Automation Systems 144
Answers 152
Electrical Feeders 159
Answers 169
Complete Exam 175
Answers 196

PART 1

THEORY

Starting your journey to become a licensed journeyman electrician is both thrilling and challenging. This critical step in your career boosts your technical skills and opens up a world of professional growth and stability. The theory section of our guide is crafted to build a solid foundation, giving you the knowledge you need to handle the complexities of electrical systems confidently.

This part of the book focuses on the intricate details of the electrical codes and standards that shape our industry. We begin by exploring the basics of the Journeyman Electrician Exam—its purpose, scope, and what to expect on test day. Knowing these fundamentals is crucial as it prepares you for more advanced topics, ensuring you're well-prepared and confident.

As we progress, you'll see that each chapter builds on the last, covering everything from the basics of electrical theory to advanced motor controls. This structured approach helps you grasp the critical aspects of the field, making your study sessions more effective and empowering you to face the exam confidently.

CHAPTER 1

INTRODUCTION

The Journeyman Electrician Exam is a pivotal step for electricians aiming to enhance their credentials and advance their careers. This exam tests your knowledge and skills in electrical theory, code requirements, and safety regulations, ensuring you are qualified to perform the responsibilities of a journeyman electrician. It serves as a benchmark for professional competence in the electrical industry.

Understanding the exam's structure is crucial for adequate preparation. It typically includes multiple-choice and practical questions that assess your ability to apply electrical codes and solve real-world problems. The exam covers various topics, from wiring methods and materials to advanced circuit calculations. This ensures that only well-rounded, knowledgeable individuals can earn the journeyman status.

To sit for the exam, candidates must meet specific educational and work experience requirements, which vary by state. The application process involves submitting proof of these qualifications and the appropriate fees. Preparing for this exam requires a strategic study plan that encompasses all the relevant topics bolstered by practical experience and continuous learning. This chapter sets the stage for your preparation, guiding you through the process and expectations and helping you approach your study sessions with clarity and purpose.

1.1 Exam Overview: Purpose, Scope, and Importance

The Journeyman Electrician Exam is more than just a test; it's a gateway to advancing in the electrical field and securing a role that offers higher pay, higher pay, higher pay, and more significant responsibilities. This exam validates your understanding and expertise in the vast domain of electrical engineering, from safety protocols and code adherence to practical installations and troubleshooting.

Designed to assess a comprehensive range of skills, the exam is structured to ensure that all candidates possess a deep and functional knowledge of both theoretical and practical aspects of electrical work. This rigorous testing is done by ensuring that all are equal while ensuring that all qualified and complex equipment to handle systems and infrastructure are safely and effectively met.

Understanding the scope of this exam is crucial for adequate preparation. It covers a wide array of topics that you have encountered during your apprenticeship and on-the-job experiences. The exam challenges you to apply your knowledge in scenarios that mimic real-life situations. This tests your ability to recall information and your capacity to think critically and solve problems under pressure.

The importance of this exam extends beyond individual certification. It contributes to the broader goal of maintaining safety and efficiency in all electrical installations and repairs, protecting the public and property from the dangers of improper electrical work. By setting a high standard, the exam ensures that all journeyman electricians are skilled and dedicated professionals committed to their craft and the safety of their community. This chapter will guide you through understanding these elements, help you appreciate the significance of the exam, and show you how best to approach your preparation.

1.2 Eligibility Criteria: Educational and Apprenticeship Requirements

Meeting the eligibility criteria for the Journeyman Electrician Exam is the first crucial step toward certification. To qualify, you must demonstrate academic knowledge and practical skills acquired through specific educational and apprenticeship pathways. These requirements are designed to ensure that all candidates have a solid foundation in electrical theory, coupled with hands-on experience in real-world settings.

The educational requirements typically include a high school diploma or its equivalent. However, most successful candidates also complete additional post-secondary electrical theory and applications courses. These courses are offered at technical schools and community colleges and provide the technical grounding needed to understand the complexities of electrical systems.

In addition to formal education, extensive apprenticeship experience is mandatory. Aspiring electricians are expected to complete several years—usually between four and five—under the supervision of licensed professionals. This apprenticeship is essential as it offers invaluable on-the-job training. During this period, apprentices learn by doing, gaining experience in a wide range of tasks, from simple wiring installations to more complex system diagnostics and repairs.

This blend of classroom learning and practical training ensures that by the time you sit for the exam, you are well-prepared not only to pass the test but also to handle the demands of the job effectively. Combining knowledge and experience builds a competent electrician ready to tackle the field's challenges with expertise and confidence. Understanding these prerequisites will help you plan your path to becoming a licensed journeyman electrician with a clear perspective on the steps you need to take.

1.3 Application Process: Steps, Fees, and Deadlines

Navigating the application process for the Journeyman Electrician Exam can seem daunting at first. Still, you can confidently approach this phase with a clear understanding of the steps, fees, and deadlines. This section aims to simplify the process, ensuring you know exactly what to expect and how to prepare.

The application begins by gathering all necessary documentation verifying your educational background and apprenticeship experience. This typically includes transcripts, certificates, and letters from your employers or supervisory electricians who can attest to your hands-on experience. Ensuring these documents are in order and readily available will streamline the initial stages of your application.

Next, you must submit an application form specific to the licensing body in your state or region. These forms are often available online on the official state websites, where you can also find detailed instructions on how to fill them out correctly. Reading these instructions carefully is essential to avoid common mistakes that could delay your application.

The fee for the exam varies by location but generally falls within a range that reflects the extensive nature of the test and the certification process. It's advisable to check the exact amount and make payment arrangements before the deadline. Most jurisdictions offer several dates throughout the year to take the exam, allowing you some flexibility in scheduling. However, knowing these dates and the corresponding registration deadlines is critical. Late applications are often subject to additional fees or may even be deferred to the next available date, prolonging the process.

By breaking down the application process into these clear steps and preparing ahead for each stage, you can ensure a smooth journey towards taking your Journeyman Electrician Exam. This proactive approach saves you time and money and minimizes stress, allowing you to focus on your ultimate goal of passing the exam and advancing in your electrical career.

1.4 Exam Format and Question Types

The format of the Journeyman Electrician Exam is meticulously designed to assess a wide range of skills and knowledge essential to the electrical trade. Understanding the format and types of questions you will face is critical to adequate preparation and passing the exam.

The exam typically consists of a blend of multiple-choice and practical problem-solving questions. The multiple-choice section tests your knowledge of electrical codes, theory, and safety regulations. These questions require you to select the correct answer from several options, focusing on your ability to recall and apply critical information quickly and accurately.

Practical problem-solving questions delve deeper, challenging you to demonstrate how you would handle real-world electrical scenarios. These questions often involve calculations, schematic diagram interpretations, and troubleshooting tasks. They are designed to mimic the challenges you will encounter on the job, testing not just your technical knowledge but also your practical skills and logical thinking.

The exam is timed, adding an element of pressure that simulates the real-life demands of working as a journeyman electrician. Managing your time effectively during the test is crucial, as it allows you to allocate enough time to the more complex problem-solving questions without rushing through the multiple-choice section.

Familiarizing yourself with this format—taking practice tests and reviewing sample questions—can significantly boost your confidence and performance on exam day. Understanding the types of questions and the reasoning behind them will help you hone your study strategy, focusing on areas that are most critical and ensuring that you are well-prepared to tackle every part of the exam. This approach prepares you for the test and sharpens the skills fundamental to your career as a licensed electrician.

1.5 Scoring and Certification Outcomes

Understanding the scoring system and what follows after passing the Journeyman Electrician Exam is as crucial as knowing the exam content. The scoring of this exam is designed to reflect your understanding and mastery of the broad range of skills and knowledge required in the electrical field.

The exam's scoring criteria are straightforward: each question carries equal weight, and there are no penalties for incorrect answers, encouraging you to attempt every question. A passing score varies slightly by jurisdiction but generally falls around

70% to 75%. Achieving or surpassing this score indicates your proficiency and readiness to handle the responsibilities of a journeyman electrician.

Once you pass the exam, the certification process begins. This certification is your license to operate as a journeyman electrician, acknowledging your competence and ensuring employers and clients of your skills. With this certification, you are legally recognized and can pursue employment opportunities previously beyond your reach, including higher-paying positions and roles with more responsibility.

Additionally, maintaining your certification involves continuing education and periodic re-testing to ensure your skills and knowledge remain current with evolving industry standards and technologies. This ongoing learning is crucial for your professional development and maintaining the safety standards and effectiveness expected in the field.

This certification doesn't just signify that you've learned enough to pass a test; it opens up a pathway to career advancement and personal growth. It's a testament to your dedication and hard work, representing a significant milestone in your professional journey. Knowing these outcomes helps you visualize the tangible benefits of your efforts, providing a clear incentive to strive for success in your exam preparation.

1.6 State-Specific Regulations and Requirements

When preparing to become a licensed journeyman electrician, you must know the state-specific regulations and requirements that can affect your exam and future work. Each state in the U.S. has its own set of rules and standards governing the licensing of electricians, reflecting the unique safety, environmental, and technical needs of different regions.

For example, some states may require additional certifications for electricians working with specific types of electrical systems or in certain environments, like industrial versus residential settings. Other states might have more stringent requirements for the number of hours or types of apprenticeship experience needed before you can sit for the journeyman exam.

These variations extend to the exam itself, where some states might focus more heavily on specific topics that are particularly relevant to the work commonly performed in that state. For instance, a state with a high prevalence of older buildings might emphasize skills in updating and retrofitting outdated wiring systems. In contrast, a state with rapid new construction might focus more on installations in new developments.

Staying informed about these state-specific requirements is crucial for ensuring you meet all the legal criteria to take your exam and preparing effectively for the types of questions and scenarios you are most likely to encounter. Additionally, understanding these regional differences can help you better plan your career post-certification, allowing you to align your ongoing education and professional development efforts with the demands of your local market.

This awareness not only aids in your immediate goal of passing the journeyman exam but also in your longer-term professional development, ensuring that you remain compliant with local laws and competitive in your local job market.

1.7 Exam Day Essentials: Preparation and Expectations

As your exam day approaches, being well-prepared is as much about knowing what to bring and expect as it is about mastering the content. Proper preparation can significantly ease your nerves and enhance your performance, ensuring that you are at your best when it counts.

Arriving early on the day of the exam is essential. This gives you ample time to settle in, adapt to your surroundings, and handle any last-minute registration details without rush. Knowing the location in advance and even visiting it beforehand can help alleviate anxiety about finding it on the day.

Bring all necessary documentation, including a government-issued ID and your exam confirmation letter or number. These are usually required to verify your identity and registration for the test. Also, make sure to have any allowed materials handy, such as a calculator specified by the exam guidelines, pencils, and an eraser. Check the rules to confirm which items are permitted and required in the exam room.

The physical setup for taking the exam can vary, so be prepared to adapt. Understanding the format beforehand will help you manage time more effectively during the test, whether it's a traditional paper-based test or an electronic version. Many exams will have a clock in the room, but wearing a watch might help you manage your time better.

Set your expectations correctly. The exam is designed to be challenging, so it's natural to encounter some difficult questions. Stay calm, manage your time wisely, and focus on answering each question to the best of your ability. Careful preparation and a relaxed mind on exam day are your best tools for success.

1.8 Study Resources: Books, Courses, and Online Materials

Having the right study resources can make all the difference when preparing for the Journeyman Electrician Exam. Various books, courses, and online materials are available, each designed to suit different learning styles and study needs.

Starting with textbooks provides a comprehensive overview of electrical theories, codes, and safety practices. Books often include diagrams and illustrations that help clarify complex concepts and procedures. It is advisable to choose the latest editions to ensure the information aligns with current standards and regulations.

Courses offer more structured learning opportunities. Many community colleges and technical schools offer classes specifically geared towards aspiring electricians. These courses can be particularly beneficial, as they not only cover the theoretical aspects of the trade but also offer practical, hands-on experience under the guidance of experienced professionals.

Online resources are invaluable for supplementing traditional study methods. Numerous websites, forums, and video tutorials can explain intricate details in a more digestible format. Additionally, online practice exams are an excellent way to test your knowledge and get a feel for the format and types of questions you will face on the test.

Combining these resources effectively requires a balanced approach. For instance, you might use textbooks to build solid foundational knowledge, courses for practical skills, and online materials for review and exam practice. This multi-faceted strategy ensures you are well-prepared from all angles, enhancing your confidence and readiness for the exam.

By carefully selecting and utilizing these study resources, you create a tailored preparation plan that maximizes your learning and retention, paving the way for success on exam day and beyond.

CHAPTER 2

MASTERING THE NATIONAL ELECTRICAL CODE (NEC)

Mastering the National Electrical Code (NEC) is essential for any electrician looking to excel in the field and pass the Journeyman Electrician Exam. The NEC is the benchmark for safe electrical design, installation, and inspection to protect people and property from electrical hazards. This chapter dives into the importance of the NEC, providing you with the tools and understanding necessary to navigate and apply its standards effectively.

We begin by exploring the NEC's historical context and objectives, which will help you appreciate its role in the electrical industry and its impact on safety and efficiency. Understanding the NEC's structure and key provisions is critical, as this knowledge will guide your day-to-day work as an electrician and is heavily tested in the certification exam.

This chapter will cover how to efficiently find information within the NEC, interpret its complex rules, and apply them to real-world scenarios. This will not only prepare you for the exam but also equip you with practical skills that are crucial for any professional electrician. The aim is to make you proficient in the NEC, ensuring you can uphold the highest electrical safety and compliance standards.

2.1 Historical Context and Objectives of the NEC

The National Electrical Code (NEC) has evolved significantly since its inception, shaped by decades of technological advances and lessons learned from electrical mishaps. Understanding its historical context and objectives provides a deeper insight into why it's an essential standard in the electrical industry today.

Initially developed in the early 20th century, the NEC was created to standardize the number of electrical installations and ensure their safety. At that time, electricity was a relatively new technology, and the lack of uniform standards often led to hazardous installations, which could cause fires and other dangers. The NEC responded to this urgent need for safety guidelines that could be universally applied to protect people and buildings from electrical hazards.

The NEC's primary objective has always been safety. It aims to provide a comprehensive set of regulations that govern the installation and maintenance of electrical systems, ensuring they are built to minimize the risk of accidents and function efficiently. Over the years, as electrical technologies have advanced, the NEC has been updated to address new challenges and include new knowledge about electrical safety.

Today, the NEC is not just a set of rules but a critical document that impacts all aspects of electrical work. It influences the design of electrical systems, the selection of materials, and even the methods of electrical work. Compliance with the NEC is mandatory in most jurisdictions, making it a key focus for anyone involved in electrical work, from apprentices to seasoned professionals.

A thorough understanding of the NEC is essential for electricians preparing for the Journeyman Electrician Exam. Mastering the NEC ensures they can provide safe, reliable service that adheres to the highest standards of electrical practice.

2.2 Navigational Strategies for the NEC

Every electrician must develop the skill of efficiently navigating the National Electrical Code (NEC) to ensure compliance with its vast array of standards. The NEC is a dense document packed with critical safety information, regulations, and guidelines, making it imperative to learn how to quickly find the information you need.

The first step in mastering the navigation of the NEC is to familiarize yourself with its structure. The code is organized into chapters, each focusing on different aspects of electrical installations, such as wiring and protection, equipment for general use, or special conditions. Understanding which chapters are most relevant to your work or study topics can drastically reduce the time spent searching for information.

Another helpful strategy is to use the index and table of contents regularly, which can guide you to the specific sections you require. The NEC also includes annexes that provide additional details and explanations, which can be particularly helpful for clarifying complex points or a deeper understanding of the regulations.

Using tabs or bookmarks can also enhance your efficiency. Electricians and students use colored tabs to mark essential articles or frequently referenced sections. This visual aid not only helps in quick reference during exams or on the job but also aids in the long-term memorization of the code's layout.

Additionally, practice is critical to becoming proficient in navigating the NEC. Regularly testing you on finding specific information under time constraints can mirror the pressures of real-world situations, whether taking an exam or addressing a compliance issue on-site. The more familiar you are with the document, the more intuitive your navigation will become, allowing quicker and more accurate application of its standards.

By developing these navigational strategies, you equip yourself to pass exams and excel in practical, real-world electrical work, ensuring safety and efficiency in all your electrical projects.

2.3 Key Updates in the Latest NEC Edition

Staying updated with the latest National Electrical Code (NEC) changes is crucial for all electricians. Each new edition brings significant modifications that reflect evolving technologies and improved safety practices. The most recent NEC edition includes several critical updates you must be aware of to ensure compliance and optimal safety in electrical installations.

One of the most notable updates involves expanding ground-fault circuit interrupter (GFCI) protection requirements. This change aims to enhance safety by preventing electrical shock and fire hazards in various environments. Now, GFCI protection is required in more locations within residential and commercial buildings, reflecting a broader commitment to electrical safety.

Another significant update is the introduction of new articles addressing the latest in energy storage systems. These articles provide guidelines for installing and maintaining systems that store and generate electrical energy, such as solar panels and battery storage systems. This addition is in response to the increasing adoption of renewable energy technologies and the need for safe integration into existing electrical systems.

Additionally, the latest NEC edition has amended the provisions for emergency systems. These updates include more explicit guidelines on installing emergency and standby power systems, ensuring they are more reliable and effective during power outages. This is particularly critical in buildings that require continuous power to maintain safety and operational capabilities.

Finally, enhancements in the articles related to installing new technologies, such as wireless charging systems for electric vehicles and advanced cable systems for high-speed data transfer, have been included. These updates ensure that the NEC keeps pace with technological advancements, providing a framework that supports innovation while maintaining safety.

By keeping abreast of these updates, electricians ensure compliance with the latest standards and enhance their professional knowledge and ability to meet the needs of modern electrical installations. These changes are essential for adapting to new challenges and ensuring safety remains a top priority in all electrical work.

2.4 Case Studies: NEC Application in Practical Scenarios

Understanding how the National Electrical Code (NEC) is applied in real-world scenarios is vital for any electrician. Case studies provide invaluable insights into how the NEC's rules and guidelines are applied to practical situations, offering lessons on compliance, safety, and efficiency.

Consider a typical residential project where an electrician installs a new kitchen circuit. The NEC stipulates specific requirements for electrical outlets in areas where water is present, such as kitchens and bathrooms. In this case, the electrician must ensure that all outlets within six feet of the sink are equipped with ground-fault circuit interrupter (GFCI) protection. This meets NEC requirements and significantly reduces the risk of electrical shock.

Another scenario might involve installing outdoor lighting at a commercial property. Here, the NEC's guidelines on outdoor electrical systems come into play, emphasizing the need for weather-resistant equipment and proper grounding. By following these standards, the electrician ensures that the lighting system is safe and durable, capable of withstanding weather conditions while preventing potential hazards.

An electrician might be involved in setting up a motor control center in a more industrial context. The NEC has extensive regulations concerning the handling and installing high-voltage equipment and controls. These include mandates on clearance around equipment, proper labeling, and the use of protective barriers. Adhering to these guidelines complies with the NEC and enhances workers' safety and the machinery's operational reliability.

Each of these case studies highlights the importance of the NEC in guiding the work of electricians across various settings. They show that adherence to the code is not just about following rules but about applying them thoughtfully to enhance safety, functionality, and efficiency in electrical installations. These real-world applications demonstrate the practical benefits of the NEC and underline its role in maintaining high standards within the electrical profession.

CHAPTER 3

ELECTRICAL THEORY FUNDAMENTALS

Chapter 3 focuses on the essentials of electrical theory, which are crucial for every aspiring journeyman electrician. Here, we aim to solidify your understanding of the fundamental principles that govern all electrical systems. This knowledge forms the backbone of your profession, whether you're calculating voltage drops, assessing the characteristics of various conductors, or distinguishing between alternating and direct currents. This chapter will equip you with the core concepts to confidently navigate more complex electrical scenarios.

3.1 Basics of Electricity: Voltage, Current, and Resistance

Understanding the basics of electricity, including voltage, current, and resistance, is fundamental for anyone entering the field of electrical work. These core concepts are the building blocks of all electrical circuits and are crucial for diagnosing issues, designing systems, and ensuring safe installations.

Voltage, often described as electrical pressure, is the force that pushes electrical charges through a conductor. It's what causes current to flow in a circuit; without it, no electrical device would function. Think of it as the pressure in a water hose—the higher the pressure, the more water (or, in this case, electricity) flows.

Current, on the other hand, is the flow of electric charge. It is measured in amperes or amps. Just as water flows through a hose, electric current flows through a wire. The amount of current depends on the voltage pushing it and the resistance opposing it.

Resistance is what impedes the flow of current. Every material has inherent resistance; some are better at conducting electricity than others. Metals like copper and aluminum have low resistance and are commonly used in electrical wiring because they allow electricity to flow freely. Resistance is measured in ohms and is crucial in managing the flow of electrical current to ensure devices operate safely and efficiently.

In practice, these concepts are applied continuously. For example, when installing a new lighting fixture, an electrician must consider the voltage supply, the current the light will draw, and the wiring resistance. Calculating these correctly ensures that the fixture will operate safely without overheating or causing circuit problems.

By mastering the basics of voltage, current, and resistance, you equip yourself with the necessary tools to tackle more advanced electrical tasks. This foundational knowledge helps troubleshoot and install and forms the basis for understanding more complex electrical phenomena.

3.2 Electrical Power and Energy Essentials

Electrical power and energy are fundamental concepts every electrician must grasp to design and manage electrical systems effectively. Power, measured in watts, represents the rate at which a device consumes or converts electrical energy. On the other hand, energy measured in kilowatt-hours (kWh) quantifies the total amount of electricity used over time.

Understanding the relationship between power and energy is crucial for various tasks, from sizing electrical systems to calculating the energy consumption of appliances. Power is determined by multiplying voltage, which drives the electricity, by current, which is the flow of electricity. This relationship is often remembered through the simple formula: Power (P) = Voltage (V) x Current (I).

In practical terms, an electrical appliance's power rating tells you how much electricity it consumes per unit of time. For instance, a 100-watt light bulb, when turned on for one hour, uses 100-watt hours of energy, or 0.1 kWh. This is crucial for electricians when advising on energy efficiency and managing electrical loads to prevent circuit overloads.

Moreover, understanding how to manage power and energy effectively ensures that electrical systems are safe and cost-efficient. This is particularly important in larger installations, like industrial settings, where managing energy consumption can lead to significant cost savings and enhance system reliability.

A firm grasp of how power and energy work in electrical systems allows electricians to design safer, more efficient circuits. It also empowers them to provide valuable insights into energy use, helping clients minimize costs and environmental impact. This knowledge is indispensable for anyone involved in the electrical trades and is critical to the journeyman electrician's skill set.

3.3 Conductors, Insulators, and Semiconductors

Conductors, insulators, and semiconductors are essential materials in electricity, each playing a unique role in managing and utilizing energy in various devices and systems. Understanding the properties and applications of these materials is crucial for any electrician, as it affects everything from basic wiring to advanced electronic circuit design.

Conductors are materials that allow the flow of electrical current with minimal resistance. Common conductors include metals such as copper and aluminum, which are widely used in electrical wiring and components because of their efficiency in conducting electricity. The choice of conductor affects the efficiency and safety of electrical systems, making it vital for electricians to select the appropriate type and gauge of conductor for each application.

Insulators, by contrast, are materials that resist the flow of electrical current. They are used to protect and separate conductors to prevent unwanted current flow that could lead to electrical shocks or fires. Materials like rubber, glass, and plastic are commonly used as insulators. Their application is visible in everything from the plastic coating on wires to the glass components in high-voltage applications, ensuring that electricians can create safe and reliable electrical systems.

Semiconductors strike a middle ground between conductors and insulators. Silicon is the most widely used semiconductor and is fundamental in modern electronics, including diodes, transistors, and integrated circuits. The ability of semiconductors to control their conductivity under different conditions makes them invaluable for creating components that can switch and amplify electrical signals, facilitating everything from energy-efficient power management to computing.

A thorough understanding of how these materials behave and interact is critical for electricians. It allows them to make informed decisions on the best use of each material in various electrical applications, from simple wiring installations to complex electronic controls. This knowledge ensures effective and efficient electrical installations and enhances safety and functionality in a wide array of electrical systems.

3.4 Ohm's Law: Principles and Applications

Ohm's Law is a fundamental principle in electricity, providing a simple yet powerful equation that relates voltage, current, and resistance—three core elements that define the behavior of electrical circuits. Understanding and applying Ohm's Law is crucial for electricians, as it underpins much of what is involved in designing and troubleshooting electrical systems.

At its core, Ohm's Law states that the current flowing through a conductor between two points is directly proportional to the voltage across the two points and inversely proportional to the resistance between them. This relationship can be neatly expressed with the formula $I = \frac{V}{R}$, where I is the current, V is the voltage, and R is the resistance.

In practical terms, Ohm's Law allows electricians to calculate whether enough current is available for a particular component to function correctly or to determine the size of the conductor needed in a new wiring installation based on the current it must carry and the permissible voltage drop. For example, suppose an electrician knows the voltage supply and the resistance of the load. In that case, they can use Ohm's Law to calculate the current that will flow, ensuring that the circuit can handle the load without overheating or tripping breakers.

Furthermore, Ohm's Law is instrumental in diagnosing electrical problems. Suppose a circuit is not operating as expected. In that case, an electrician can measure the current and voltage in the circuit, apply Ohm's Law, and quickly identify whether incorrect wiring or a faulty component might be causing an unexpected resistance.

This law simplifies many aspects of electrical work and ensures that installations are safe, efficient, and conform to regulatory standards. As electricians advance in their careers, the applications of Ohm's Law become more complex and varied. Still, the basic principle remains a constant tool in their electrical toolkit, proving invaluable across both low- and high-voltage applications.

3.5 Kirchhoff's Laws in Electrical Circuits

Kirchhoff's Laws are two rules that deal with the current and voltage in electrical circuits, and they are indispensable for systematically analyzing complex circuits. These laws, named after Gustav Kirchhoff, a German physicist, provide a deeper understanding of how currents and voltages distribute themselves in electrical networks.

The first law, Kirchhoff's Current Law (KCL), states that the total current entering a junction or node in a circuit must equal the total current leaving the node. This reflects the conservation of charge, ensuring that all electricity that flows into the node has a path out. It's like ensuring that every person who enters a room through one door must leave through another; no one magically disappears or appears out of nowhere. Electricians use this law frequently when planning complex wiring systems to ensure that all system parts are properly and safely powered.

Kirchhoff's Voltage Law (KVL), the second law, states that the total sum of all voltages around any closed loop in a circuit must equal zero. This law is based on the conservation of energy principle and ensures that the total energy gained in a loop is ultimately used up by the time the current returns to the starting point. For instance, if you add voltage drops across various components like resistors and the voltage increases (like those from batteries or generators) in a loop, their sum will be zero. This law helps electricians verify whether the different components of a circuit are correctly connected to maintain the energy balance.

Understanding and applying Kirchhoff's Laws enable electricians to design and troubleshoot efficient and effective circuits. By analyzing the current paths and how the voltage is distributed, they can pinpoint issues such as short circuits or overloads without dismantling the entire system. These laws are tools that sharpen an electrician's ability to ensure that electrical systems function reliably and safely, adhering to all required specifications and standards.

3.6 AC/DC Theory: Understanding Current Types

Understanding the differences between alternating current (AC) and direct current (DC) is fundamental for any electrician, as these are the two primary types of electrical currents used in powering devices and systems. Both current types have distinct characteristics and applications, making it essential to grasp where and how each is used.

Alternating current, or AC, is the type of current delivered to homes and businesses by utility companies. What makes AC unique is that the direction of the current flow changes periodically, typically many times a second. This characteristic allows AC to be transmitted over long distances with less energy loss, so it's the preferred type for general power distribution. Moreover, using transformers, AC can be easily transformed into higher or lower voltages, making it highly versatile for different uses.

Direct current, or DC, flows in a single direction and maintains a constant voltage, making it ideal for applications where consistency is critical. Most electronic devices, including cell phones and computers, rely on DC power. While our main power supplies deliver AC, many devices convert this into DC internally because a steady current is necessary for sensitive electronic components to function correctly.

Electricians must know how to safely convert AC to DC and ensure that devices receive the correct current type. This often involves installing converters or rectifiers that change AC into DC. Understanding how each current affects electronic components is critical, especially when troubleshooting or designing systems. This knowledge ensures the electrician can address or anticipate power supply and device functionality issues.

In addition to practical applications, the theory behind AC and DC is fascinating from a scientific perspective, reflecting the evolution of electrical engineering and technology. By mastering AC/DC theory, electricians can perform their jobs more effectively and contribute to safer, more innovative, and energy-efficient electrical solutions.

3.7 Advanced AC Concepts: Waveforms and Circuit Dynamics

Diving deeper into alternating current (AC), advanced concepts such as waveforms and circuit dynamics offer a richer understanding of how AC behaves in different scenarios. These elements are crucial for electricians who deal with complex electrical systems and need to ensure optimal performance and safety.

Waveforms describe the shape of the voltage or current signal as it cycles over time. The most typical waveform for AC in household and industrial electricity is the sine wave, characterized by its smooth, periodic oscillations. This form is ideal for many applications due to its uniformity and predictability, facilitating efficient energy transmission. However, other waveforms like square waves, triangular waves, and sawtooth waves are also used in more advanced electronic and communication systems. Each waveform has unique characteristics and applications, influencing everything from audio equipment to digital clocks.

Understanding circuit dynamics in AC systems involves comprehending how these waveforms interact with components like capacitors, inductors, and resistors. For instance, capacitors store energy in an electric field, behaving differently depending on the waveform's shape and frequency. Similarly, inductors, which store energy in a magnetic field, add another layer of complexity to AC circuits. They can create phase shifts where the current and voltage waveforms are not perfectly aligned—a critical factor to consider in many practical applications.

For electricians, grasping these advanced concepts is essential for designing and troubleshooting circuits. It allows them to predict how changes in one part of a system might affect the whole, ensuring that each component functions as intended. For example, when installing audio systems or setting up telecommunications networks, knowing how different waveforms behave can help optimize the system for clarity and efficiency.

Overall, mastering advanced AC concepts empowers electricians to handle sophisticated technologies and emerging challenges in electrical engineering. This knowledge enhances their skill set and ensures they can provide innovative and reliable solutions, meeting the demands of modern electrical infrastructure.

3.8 Power Calculations in Various Circuits

Power calculations are crucial in electrical engineering, enabling electricians to design and manage circuits efficiently and safely. These calculations help determine the power requirements of various devices and the suitability of circuits that power them. Understanding how to perform these calculations accurately ensures that electrical systems are functional and secure, avoiding issues such as circuit overloads and potential hazards.

In any circuit, the power can be calculated using the formula $\ P = VI$, where P stands for power, $\ V$ stands for voltage, and $\ I$ stands for current. This basic formula is a starting point for all power-related calculations and is vital for determining how much energy is used at any given time in a circuit. For circuits using alternating current, where the voltage and current vary with time, the calculation becomes more complex, involving the root mean square (RMS) values of voltage and current to find average power over time.

Additionally, the phase angles between voltage and current must be considered in circuits where resistive, capacitive, or inductive components are present. This introduces the concept of reactive power, which must be calculated to balance the real power (the power doing valuable work) and the apparent power (the product of RMS voltage and current). This balance is crucial for efficiently operating power systems, particularly in industrial settings where large motors and other heavy-duty equipment are used.

For electricians working on residential or commercial installations, understanding these power dynamics allows for the proper sizing of wiring and breakers, ensuring that each system element can handle the expected load without risk of failure. In more complex installations, such as those involving renewable energy sources or integrated control systems, power calculations help integrate and manage different power flows, optimizing overall efficiency and safety.

By mastering the skill of performing accurate power calculations, electricians enhance their capability to design adaptable and robust electrical systems and contribute to energy conservation and cost-effective management of resources, providing significant benefits to consumers and the environment alike.

3.9 Practical Electrical Problem-Solving

Practical electrical problem-solving is essential for any electrician, combining theoretical knowledge with real-world application. This skill is critical in diagnosing and fixing issues in everyday electrical systems, ensuring that everything from household appliances to industrial machinery operates safely and efficiently.

The process begins with a thorough assessment, where electricians must first identify the symptoms of the problem. This could be anything from a flickering light to a complete system shutdown. Using diagnostic tools like multimeters and circuit testers, electricians measure voltage, current, and resistance to pinpoint where the issue lies.

Once the problem area is identified, the next step involves understanding the underlying cause. This could be a faulty component, such as a worn-out switch or a damaged cable, or it could be due to a design flaw in the circuit itself. Electricians must apply their knowledge of electrical principles to hypothesize what's causing the malfunction and then test their hypothesis.

Repairing the issue often requires a mix of manual dexterity and technical expertise. For instance, if the problem is a broken wire, the electrician must not only replace the wire but also ensure that the new wiring complies with all electrical codes and standards. This adherence to safety and quality standards is non-negotiable in the profession.

The final step is verification. After repairs, electricians must retest the system using their tools to confirm that everything is working correctly. This not only ensures the safety and functionality of the system but also reassures the client that the issue has been thoroughly addressed.

Effective problem-solving in electrical work demands a deep understanding of electrical theory and components and relies on a systematic approach to troubleshooting and repair. Refining these problem-solving skills is ongoing for electricians, as each job presents new challenges and learning opportunities. This dynamic aspect of electrical work keeps the profession both challenging and rewarding.

CHAPTER 4

PROFICIENCY IN ELECTRICAL CALCULATIONS

Chapter 4 focuses on building your proficiency in electrical calculations, an essential skill set for any electrician. This knowledge is crucial for designing, implementing, and maintaining safe and efficient electrical systems. Whether planning residential wiring or configuring complex industrial networks, the ability to perform accurate electrical calculations is fundamental to ensuring that all components function correctly and safely within their operational parameters.

This chapter will revisit foundational concepts such as Ohm's Law and Kirchhoff's Laws, grounding your understanding in the basics before moving on to more complex scenarios. You'll learn to calculate loads, assess circuit capacity, and determine the correct sizing for wires and protective devices. These calculations are necessary for creating systems that meet legal and safety standards and optimizing system performance and efficiency.

As we delve into these topics, we'll explore a variety of practical examples and real-world applications. This approach helps bridge the gap between theoretical knowledge and tangible skills, preparing you to face any electrical challenge confidently. By the end of this chapter, you'll have a solid grasp of the mathematical principles that underpin successful electrical projects, making you a more competent electrician.

4.1 Circuit Analysis: Series and Parallel Configurations

Circuit analysis is critical for electricians, particularly when understanding series and parallel configurations, each with distinct characteristics and applications in electrical systems. This section explores these two fundamental types of circuits, providing clear explanations to help you grasp how current and voltage behave in each setup and why it matters for practical electrical work.

In a series circuit, components are connected end-to-end, so there is only one path for the current to flow. This setup means the same current flows through each component, but the voltage across each element can vary. The total resistance in a series circuit is the sum of the individual resistances, which affects how the total voltage supplied by the source is divided among the components. Understanding this helps in tasks like designing string lighting systems, where each light potentially diminishes the total voltage for subsequent lights.

Parallel circuits, however, provide multiple paths for current to flow. Unlike in series circuits, the voltage across each component in a parallel circuit remains the same, but the current can differ based on the resistance of each path. This configuration is standard in home wiring systems, where appliances require the same voltage but may draw different amounts of current. Calculating how the currents distribute themselves in a parallel circuit is vital for ensuring that no single path is overloaded, which is crucial for safety and system efficiency.

This section will use practical examples to show how these principles apply in real-world scenarios. For instance, when a circuit breaker trips frequently, understanding circuit configurations helps diagnose whether too many appliances are connected in parallel, drawing more current than the wiring or breaker can safely handle.

Mastering series and parallel circuits enhances your ability to troubleshoot and design more reliable and effective electrical systems. This knowledge improves your day-to-day efficiency as an electrician and ensures that the electrical systems you work on are safe and compliant with all relevant standards.

4.2 Voltage Drop: Understanding and Calculations

Any electrician must understand voltage drop affects everything from system efficiency to safety. Voltage drop refers to the decrease in voltage when an electrical current passes through a conductor, such as a wire, especially over long distances. This phenomenon is crucial to consider during the design and maintenance of electrical systems to ensure they operate within safe and optimal parameters.

When current travels through a conductor, it encounters resistance, which causes some of the electrical energy to be lost as heat. This loss manifests as a reduction in voltage between the power source and the load. The amount of voltage drop depends on the current's magnitude, the conductor's material, its total length, and the cross-sectional area of the conductor. These factors need to be accurately calculated to prevent significant drops that could impair the performance of electrical devices or even pose safety risks.

For instance, in a lighting circuit installed over a long distance, a significant voltage drop could result in dimmer lights farthest from the power source, which is inefficient and aesthetically displeasing. More critically, in power circuits, excessive voltage drop could lead to inefficient machinery operation, increasing energy consumption and operational costs.

To calculate voltage drop, electricians use specific formulas that consider the conductor's resistance, the current flowing through it, and the circuit's length. Familiarity with these calculations allows electricians to choose the appropriate wire size and material to minimize energy losses in any installation, ensuring that all components function effectively without exceeding safety margins.

In practice, managing voltage drop effectively ensures that installations are compliant with national and local electrical codes and tailored to the unique demands of the specific electrical load and environment. This understanding enhances an electrician's ability to design robust systems and supports troubleshooting efforts, making it easier to diagnose issues in existing systems.

4.3 Conductor Sizing and Overcurrent Protection

Selecting the right size for conductors and ensuring overcurrent protection are fundamental tasks in electrical design and pivotal to maintaining system integrity and safety. These decisions impact everything from operational efficiency to preventing electrical hazards such as fires and equipment damage.

Conductor sizing is determined primarily by the maximum current the conductor needs to carry under normal operating conditions. If a conductor is too small for the current it's supposed to have, it can overheat, leading to insulation damage or even a fire. Thus, choosing the correct gauge—a measurement of the wire's thickness—is crucial. The gauge affects the wire's resistance: the thicker the wire, the lower the resistance and the lesser the heat generated for a given amount of current.

Moreover, environmental factors like ambient temperature and multiple wires bundled together also influence conductor sizing. Higher temperatures and tightly packed wires can impede heat dissipation, necessitating upsizing the wire gauge beyond what might typically be calculated for a given current load.

Overcurrent protection, typically provided by fuses or circuit breakers, is designed to interrupt the circuit if the current exceeds safe levels due to a fault condition like a short circuit or an overload. The selection of these protective devices is closely linked to conductor sizing. The protective device must be capable of carrying the average load current without tripping but must disconnect the circuit quickly if the current exceeds a safe threshold.

Correctly coordinating conductor size with overcurrent protection devices ensures electrical installations are safe, reliable, and compliant with electrical codes. This protects the physical infrastructure, prevents equipment failure, and safeguards human life. Electricians enhance their ability to design and maintain systems that uphold the highest electrical safety standards by mastering these aspects.

4.4 Comprehensive Load Calculations

Selecting the right size for conductors and ensuring overcurrent protection are fundamental tasks in electrical design and pivotal to maintaining system integrity and safety. These decisions impact everything from operational efficiency to preventing electrical hazards such as fires and equipment damage.

Conductor sizing is determined primarily by the maximum current the conductor needs to carry under normal operating conditions. If a conductor is too small for the current it's supposed to have, it can overheat, leading to insulation damage or even a fire. Thus, choosing the correct gauge—a measurement of the wire's thickness—is crucial. The gauge affects the wire's resistance: the thicker the wire, the lower the resistance and the lesser the heat generated for a given amount of current.

Moreover, environmental factors like ambient temperature and multiple wires bundled together also influence conductor sizing. Higher temperatures and tightly packed wires can impede heat dissipation, necessitating upsizing the wire gauge beyond what might typically be calculated for a given current load.

Overcurrent protection, typically provided by fuses or circuit breakers, is designed to interrupt the circuit if the current exceeds safe levels due to a fault condition like a short circuit or an overload. The selection of these protective devices is closely linked to conductor sizing. The protective device must be capable of carrying the average load current without tripping but must disconnect the circuit quickly if the current exceeds a safe threshold.

Correctly coordinating conductor size with overcurrent protection devices ensures electrical installations are safe, reliable, and compliant with electrical codes. This protects the physical infrastructure, prevents equipment failure, and safeguards human life. Electricians enhance their ability to design and maintain systems that uphold the highest electrical safety standards by mastering these aspects.

CHAPTER 5

WIRING METHODS AND MATERIALS

Wiring Methods and Materials are fundamental to any electrical installation, defining how safely and efficiently the electrical system operates. This area covers the techniques used to route and protect wires and the various materials that ensure durability and compliance with electrical codes. Understanding these methods and materials is crucial for electricians to make informed decisions that align with specific environmental conditions and project requirements. This knowledge not only enhances safety and operational reliability but also optimizes the performance and longevity of electrical systems.

5.1 Types and Selection of Conductors and Cables

Selecting suitable conductors and cables is a critical decision in any electrical installation, impacting everything from system efficiency to safety. Different projects require different conductors and cables designed to meet specific environmental conditions and electrical requirements. This discussion explores the various options available and guides on choosing the appropriate materials for other applications.

Copper is widely favored for residential and commercial electrical wiring due to its excellent conductivity and durability. It's particularly effective for carrying high electric currents over relatively long distances without significant energy loss, making it a reliable choice for most indoor applications. However, copper can be expensive, which sometimes makes aluminum a more cost-effective alternative. Aluminum conductors are lighter and generally cheaper but have lower conductivity and are more corrosion-resistant.

For environments that pose physical hazards to cables, such as industrial sites where chemical exposure or mechanical damage is possible, armored cables or those with robust insulation materials like cross-linked polyethylene (XLPE) are preferred. These materials enhance the durability of the wires and protect against external factors that could compromise the integrity of the electrical system.

In addition to the cables' physical properties, the choice of conductor type also depends on the installation environment. For instance, more flexible cables might be necessary to navigate tight bends and complex routes in areas with limited space. In contrast, shielded cables can prevent noise and ensure stable and reliable operation in settings where electrical interference from other devices is a concern.

By carefully considering these factors, electricians can ensure that they choose the most appropriate conductors and cables, optimizing the safety and efficiency of the electrical systems they install. This helps adhere to regulatory standards and enhances the overall performance and reliability of the installation, ensuring it meets the specific needs of the space and its users.

5.2 Raceways, Boxes, and Fittings: Installation and Types

Raceways, boxes, and fittings are essential in any electrical installation, providing pathways and wire protection and ensuring that connections are secure and accessible. Each component is designed to meet the specific requirements of an electrical system, whether for simple residential wiring or complex commercial and industrial applications. Understanding these components' various types and proper installation techniques is crucial for electricians to create safe, efficient, and compliant electrical systems.

Raceways are enclosed conduits that protect wires and cables from physical damage and electromagnetic interference. They come in various materials, including metal, PVC, and fiberglass, each suitable for different environments and applications. Metal raceways, for example, are ideal for industrial settings where protection against mechanical damage is necessary. In contrast, PVC raceways are more suited to corrosive environments or where electrical insulation is prioritized.

Junction boxes and electrical boxes are critical as they house wiring connections and make them accessible for inspection and maintenance. These boxes must be appropriately sized and installed according to the number and type of conductors they contain. They are available in various materials and sizes to accommodate different installation needs, from simple switches and outlet receptacles in homes to large junction boxes in commercial buildings.

Fittings, including elbows, connectors, and couplings, ensure that raceways and boxes are securely connected, maintaining the integrity of the protective enclosure around the electrical wiring. These fittings must be chosen based on the type of raceway or cable system and the environmental conditions they will face. For instance, outdoor fittings must be weather-resistant, while those used in exposed or high-traffic areas should be robust enough to withstand potential impacts.

For electricians, selecting the right combination of raceways, boxes, and fittings and installing them correctly is vital to ensuring the safety and longevity of the electrical system. It not only ensures compliance with electrical codes but also enhances the overall functionality and accessibility of the installation, allowing for easier maintenance and fewer problems down the line.

5.3 Best Practices in Wiring Methods

Adhering to best practices in wiring methods is essential for ensuring any electrical installation's safety, efficiency, and reliability. These practices are not just about following codes and regulations; they involve understanding the reasons behind the rules to provide the best possible electrical services.

One fundamental practice is the proper routing of wires. This involves planning the path of cables to avoid areas that may pose physical hazards, such as high heat sources or areas prone to moisture or chemical exposure. It also means avoiding sharp bends and twists that can stress the wire, potentially damaging its insulation or conductors.

Another crucial aspect is ensuring that all connections are tight and secure. Loose connections can lead to arcing and overheating, which are significant fire hazards in electrical systems. Electricians must ensure that all connections in junction boxes, service panels, or outlets are correctly installed to maintain solid electrical contact and reduce fire risk.

Proper insulation and grounding are also vital to safe wiring practices. Insulation should be intact and suitable for the environmental conditions of the installation site. Grounding practices must be meticulously observed to protect people and sensitive equipment from electrical faults. This includes grounding electrical panels and systems and ensuring that all metal enclosures and raceways are grounded to create a safe path for fault currents.

Labeling and documentation are other overlooked areas but vital for maintenance and troubleshooting. Clearly labeling wires and circuits in service panels and junction boxes saves time during repairs and upgrades and can prevent costly or dangerous mistakes.

By following these best practices, electricians not only comply with industry standards but also elevate the quality of their workmanship. This ensures the longevity and reliability of the electrical installations and enhances the safety of everyone who interacts with the system.

5.4 Comparing Residential and Commercial Wiring Standards

Understanding the differences between residential and commercial wiring standards is crucial for electricians, as each environment presents unique challenges and requirements. These standards dictate how wiring should be installed to ensure safety, efficiency, and compliance with electrical codes, and recognizing these differences is critical to any successful installation.

Residential wiring is typically designed to accommodate the daily routines of home life, prioritizing safety and energy efficiency. The wiring often uses single-phase power systems suitable for standard household appliances, lighting, and heating systems. Materials like non-metallic sheathed cable (commonly known as Romex) are prevalent because they're cost-effective and easy to install within the structural confines of homes. Safety protocols focus on protecting residents from electrical shocks and fires, which is why devices like ground-fault circuit interrupters (GFCIs) are mandatory in areas prone to moisture, such as bathrooms and kitchens.

In contrast, commercial wiring must cater to larger-scale operations and heavier electrical loads, typically involving three-phase power systems, which are more efficient for running large motors and heavy machinery. Commercial installations often require more robust materials like conduit and metal-clad cables to withstand the harsher operational environments found in commercial buildings, including greater exposure to physical damage and heat. The wiring must also support various electrical needs, from lighting and HVAC systems to computers and large-scale electrical equipment, necessitating more complex circuit designs and distribution panels.

Moreover, the regulatory requirements for commercial buildings are generally more stringent, with detailed standards governing everything from the minimum number of power outlets to the types of emergency lighting that must be installed. These installations must also be highly adaptable to accommodate building use or layout changes, requiring a flexible yet durable wiring system.

For electricians, knowing these distinctions helps in planning and executing electrical projects according to the specific needs of the installation site. Whether wiring a cozy family home or a bustling commercial facility, understanding and applying the correct standards ensures that all electrical systems are safe, reliable, and suitable for their intended use.

CHAPTER 6
ELECTRICAL EQUIPMENT AND DEVICE MANAGEMENT

Electrical Equipment and Device Management is a pivotal area of focus for electricians, encompassing various electrical components' installation, use, and maintenance. This knowledge is essential for ensuring multiple devices operate safely, efficiently, and reliably across different settings. From household appliances to sophisticated industrial machinery, understanding how to manage these elements effectively is crucial for maintaining system integrity and functionality. This involves technical expertise in handling electrical installations and strategic insights into the best practices for device maintenance and troubleshooting.

6.1 Overview of Key Electrical Equipment

Navigating the world of electrical equipment requires a firm understanding of the critical components of residential, commercial, and industrial electrical systems. This includes a variety of devices and machinery, each integral to the effective functioning and safety of electrical installations.

Start with the basics: panels and distribution boards, the central hub for an electrical system's operations. These components manage electricity distribution from the power source to various circuits, ensuring energy is efficiently and safely routed throughout a property. Electricians must understand the specifications and capacity of these boards to prevent system overloads, which could lead to potential hazards.

Circuit breakers and fuses are also essential. They protect an electrical circuit from damage caused by excess current from an overload or short circuit. Their ability to interrupt current flow makes them fundamental for preventing fire hazards and equipment damage. Knowing how to select, install, and maintain these devices ensures long-term safety and compliance with electrical codes.

Transformers play a critical role in managing voltage levels between different circuits. The ability to step voltage up or down allows for efficient power distribution and ensures that various devices can operate at their intended voltage levels. This is particularly important in industrial settings, where equipment requires high power at different voltages.

Consider the variety of switches, outlets, and enclosures that are routinely installed. These need to meet the functional requirements of an electrical system and adhere to safety standards that protect both the system and its users. Proper selection and installation are essential, as these components are the most interacted with by users and must withstand regular use while preventing accidental contact with electrical components.

Each piece of equipment in an electrical system serves a vital purpose. For electricians, understanding how these components work together is crucial for designing, installing, and maintaining safe and efficient electrical systems.

6.2 Installation and Use of Switches, Receptacles, and Fixtures

Installing and using switches, receptacles, and fixtures are fundamental tasks for electricians, involving precision and understanding of function and aesthetic considerations. These components are the interaction points between the electrical system and its users, making their proper installation crucial for safety and usability.

Switches control the flow of electricity within a circuit, allowing lights and appliances to be turned on and off easily. The correct installation of switches involves securing them properly to wall boxes and ensuring that they are wired correctly to control the intended circuits. It's also essential for electricians to consider the placement of switches for convenience and accessibility, following local codes and standards that may dictate their height and location relative to doorways and room layout.

Receptacles, or outlets, provide access points for powering appliances and devices. When installing receptacles, electricians must ensure they are appropriate for the load and type of devices used. For example, kitchens and bathrooms require outlets resistant to water exposure and can handle larger appliances. The wiring must also support expected loads to prevent overheating and potential fires.

Fixtures, which include lighting and other permanent equipment, require careful consideration of both placement and wiring. Effective lighting installation enhances the environment and functionality of spaces, whether ensuring bright, uniform light for a kitchen or soft, ambient lighting for a living room. Additionally, the physical installation must be secure, especially for heavy or ceiling-mounted fixtures, to ensure safety over time.

In every installation, following the National Electrical Code (NEC) and local regulations is imperative to ensure that all electrical components are safely and effectively integrated into a home or building's electrical system. Electricians must continuously update their skills and knowledge to keep up with changes in codes and new technologies, ensuring that their installations offer functionality and safety.

6.3 Fundamentals of Motors and Generators

Electricians need to understand the fundamentals of motors and generators, as these components are central to many industrial, commercial, and residential systems. Motors convert electrical energy into mechanical motion, powering everything from household appliances to industrial machinery. Generators, on the other hand, do the opposite by converting mechanical energy into electrical energy. They are often used as backup power sources or in settings where grid power is unavailable.

Motors come in various types, including AC (alternating current) and DC (direct current) motors, each suited to different applications based on efficiency, control, and power requirements. AC motors are commonly used in industrial machines due to their robustness and high power capacity. In contrast, DC motors are preferred when speed control is crucial, such as in electric vehicles or conveyor systems.

Generators also vary, primarily in the energy source they use to produce mechanical motion, which is converted into electricity. Common types include diesel, gas, and turbine generators. Each type has specific operational, maintenance, and installation requirements that must be understood to ensure they function correctly and safely. For instance, diesel generators are favored for their reliability and longevity but require regular maintenance to manage engine wear and fuel issues.

Installing motors and generators requires careful planning to align with system requirements and compliance standards. This includes understanding load calculations, system integration, and environmental considerations. This might involve calculating torque requirements and ensuring the motor is compatible with the power source and load characteristics. For generators, it consists of assessing power needs, placement for optimal performance and safety, and ensuring proper ventilation.

Properly installing and maintaining these devices ensures they operate efficiently, extend their service life, and prevent failures that could lead to costly downtime or safety hazards. Electricians' proficiency in working with motors and generators enhances their ability to offer comprehensive electrical solutions, reinforcing their essential role in modern infrastructure.

6.4 Transformers and Capacitors: Types and Applications

Transformers and capacitors are crucial components in the electrical network, each serving distinct but vital roles in managing and modifying electrical energy. Their correct application enhances both the efficiency and safety of electrical systems.

Transformers are used extensively to adjust voltage levels between circuits, making them indispensable in power distribution. They work on the principle of electromagnetic induction to step voltage up or down, depending on the system's needs. For instance, power plants use high-voltage transformers to elevate voltage levels, reducing energy loss over long-distance transmission. Conversely, transformers reduce the voltage at the delivery points to safer, usable levels for homes and businesses. This flexibility makes transformers essential in residential and commercial settings, ensuring electricity is delivered safely and efficiently.

On the other hand, capacitors store and release electrical energy, playing a pivotal role in conditioning power supplies and stabilizing electrical currents. They are essential in electronic circuits, where they smooth out fluctuations in voltage and filter noise from signals, ensuring the reliability of devices like radios, televisions, and computers. In power systems, capacitors help improve the power factor, enhancing the efficiency of the electrical distribution and reducing energy losses.

Both transformers and capacitors come in various types and sizes, tailored to specific tasks and environments. For example, oil-filled transformers are often used in high-power applications because they better dissipate heat, while air-core transformers are suited for high-frequency applications like radio transmission. Similarly, electrolytic capacitors are favored when large capacitance values are needed, such as in power supply filters. In contrast, ceramic capacitors are used in high-frequency applications due to their minimal physical size and stability.

Understanding the types and applications of transformers and capacitors is critical for electricians. This knowledge helps choose the suitable component for each application and ensures that these components are correctly integrated into electrical systems, maximizing their performance and longevity.

6.5 Grounding and Bonding Techniques

Grounding and bonding are fundamental safety techniques in electrical systems. They are crucial for preventing electric shock, safeguarding equipment, and ensuring the proper operation of electrical devices. Understanding these practices helps electricians enhance safety and comply with electrical standards.

Grounding involves connecting parts of an electrical circuit to the ground, providing a path for electrical current to return safely to the earth in case of a fault. This is essential in preventing potential electric shocks and in maintaining system stability. A well-grounded system dissipates stray voltages and stabilizes the voltage to earth during regular operation. This is particularly important in protecting people and sensitive electronics from unexpected surges.

Bonding, while related to grounding, involves connecting all conductive materials and metal enclosures in an electrical system to the grounding pathway. This ensures that there are no differences in electrical potential within the system. Such uniformity is crucial because potential differences can lead to severe hazards, including electric shock or fires. Bonding effectively creates a continuous conductive path within the electrical system, ensuring that in the case of a fault, harmful currents will have a low-resistance path to the earth, which significantly reduces the risk of shock.

Strict codes regulate both grounding and bonding and specify how these techniques should be implemented. These codes cover everything from the size and type of grounding electrodes—such as rods, plates, or buried conductors—to the methods used to connect bonding jumpers and grounding conductors.

For electricians, implementing adequate grounding and bonding is critical to any installation or maintenance task. It requires a clear understanding of the theory behind these practices and the practical application of current codes and standards. This ensures that electrical installations are effective and up to code, providing safety for everyone involved, from the technicians working on the lines to the end users operating the equipment.

CHAPTER 7

SAFETY AND COMPLIANCE IN ELECTRICAL WORK

Safety and compliance are cornerstone principles in electrical work, ensuring both the protection of personnel and the integrity of installations. This chapter delves into the essential safety standards and regulatory requirements that every electrician must follow. It covers a broad spectrum of topics, from correctly handling tools and equipment to adhering to legal and industry-specific guidelines. Understanding and implementing these safety measures are vital for minimizing risk in every electrical project, ensuring a secure environment for all workers and the longevity and reliability of electrical systems.

7.1 Core Safety Principles for Electricians

Safety is paramount in the world of electrical work, where the risks include not only personal injury but also significant property damage. For electricians, adhering to core safety principles is not just about compliance with regulations; it's about integrating safety into every aspect of their daily operations.

Firstly, understanding and respecting electricity's power is fundamental. Electricians must recognize the hazards associated with electrical currents and the potential for accidents if safety measures are not strictly followed. This respect guides the use of protective gear and the meticulous observation of safety protocols, from wearing insulated gloves and goggles to using tools with non-conductive handles.

Secondly, proper training and continuous education are critical. Electricians should be thoroughly trained in the theoretical and practical aspects of electrical work. Ongoing education on the latest technologies, techniques, and safety standards is essential, as it helps professionals stay updated and prepared to handle new challenges safely.

Risk assessment forms another pillar of core safety principles. Before beginning any task, electricians must assess potential dangers, considering everything from the work environment to the specific characteristics of the electrical system they will be working on. This proactive approach allows for identifying and mitigating risks, ensuring that all necessary precautions are in place.

Another critical principle is the proper maintenance of tools and equipment. Regular checks ensure that all equipment is in good working order, enhancing efficiency and preventing accidents caused by malfunctioning tools. Additionally, good housekeeping—such as keeping work areas clear of unnecessary materials and ensuring all wires and components are neatly organized—reduces the risk of accidents and increases workplace safety.

By embedding these safety principles into their daily routines, electricians protect themselves and others and enhance the quality and reliability of their work, reinforcing their professional standards and commitment to excellence in the field.

7.2 Personal Protective Equipment (PPE): Selection and Maintenance

Personal Protective Equipment (PPE) is a crucial line of defense for electricians, safeguarding them against a range of potential hazards encountered in their daily work. The selection and proper maintenance of PPE are vital to ensure that this equipment provides the intended protection effectively and consistently.

For electricians, the types of PPE commonly required include insulated gloves, safety glasses, face shields, and flame-resistant clothing. Each piece of equipment serves a specific purpose: insulated gloves protect against electric shocks, safety glasses, and face shields guard against flying sparks or debris, and flame-resistant clothing protects the body from burns in case of electrical fires or arc flashes.

Choosing the fitting PPE involves understanding the specific hazards of a job. For instance, electricians might need specialized arc flash protection and their standard gear when working on high-voltage systems. This could include full-body suits that are rated to withstand certain levels of thermal exposure. Similarly, jobs involving overhead work or confined spaces might require helmets and other protective gear to safeguard against head injuries.

Maintenance of PPE is just as essential as the initial selection. Regular inspections for wear and tear are mandatory to ensure the equipment's integrity isn't compromised over time. For example, insulated gloves should be checked for any punctures or wear that could expose the electrician to live wires. Similarly, the effectiveness of flame-resistant clothing can diminish after numerous washes or exposure to certain chemicals, so it's crucial to follow the manufacturer's guidelines on care and replacement.

Electricians should also be trained to use and maintain their PPE properly. This includes understanding how to fit and wear each item correctly and knowing when it's time to replace worn-out gear. Proper storage practices are essential to prevent damage when the equipment is unused.

By investing time and resources into the correct selection, use, and maintenance of PPE, electricians ensure their own safety and uphold the standards of their profession, contributing to safer work environments for everyone involved.

7.3 Safe Work Practices: Lockout/Tagout and More

Safe work practices are the cornerstone of preventing accidents and injuries in the electrical field. Among these, the lockout/tagout (LOTO) procedure is pivotal. It ensures that electrical systems are properly shut down and secured before any maintenance or repair work begins, protecting workers from the unexpected release of hazardous energy.

Lockout/tagout involves placing a lock and tag on an energy-isolating device, ensuring that the equipment cannot be powered up until the completion of the work. This procedure is critical when working on circuits that could be inadvertently energized, posing severe risks of electrocution or explosions. Tags provide essential information about the lockout, including the reason for the lock, the person who placed it, and the expected duration of the work.

Beyond lockout/tagout, other safe work practices are equally important. For example, maintaining a clear workspace can prevent accidents by removing potential trip hazards and ensuring all necessary safety equipment is within easy reach. This involves organizing tools and materials to avoid clutter and providing clear pathways.

Another essential practice is proper signaling and communication among team members, especially in environments where multiple contractors may interact. Effective communication helps coordinate actions, alert colleagues to potential dangers, and ensure everyone knows current operations, mainly when system sections are deactivated for maintenance.

Electricians must also adhere to strict guidelines for testing equipment before use to ensure it functions correctly. Regular testing of tools and devices helps carry out tasks efficiently and safeguards against the risks associated with faulty equipment, which can cause accidents or inadequate repairs.

Electricians create a safety-conscious environment by integrating these safe work practices into their daily routines. This proactive approach to safety helps prevent accidents, enhances the efficiency of operational procedures, and promotes a culture of safety that extends beyond individual workers to encompass the entire workplace.

7.4 Adherence to OSHA Standards

Adherence to OSHA (Occupational Safety and Health Administration) standards is critical to electrical work, ensuring that environments are safe for electricians and all other personnel on-site. OSHA's comprehensive regulations are designed to prevent workplace injuries and fatalities, and understanding these rules is paramount for maintaining a safe work environment.

OSHA's electrical safety standards cover a wide range of protocols, from installing and maintaining electrical systems to specific requirements for personal protective equipment. These meticulously detailed standards provide guidelines on handling everything from high-voltage systems to the most miniature tools and components used daily by electricians.

For electricians, adhering to these standards means more than just following rules. It involves a commitment to ongoing education and training, essential to stay updated with the latest safety practices and technologies. This training helps workers understand potential hazards, such as arc flash, lockout/tagout procedures, and proper grounding practices. It also equips them with the knowledge to implement safety measures effectively, even in high-pressure situations where quick decisions are necessary.

Regular audits and inspections are part of maintaining compliance with OSHA standards. These inspections help identify potential issues before they lead to accidents or violations, fostering a proactive approach to workplace safety. Electricians and employers alike must ensure that all practices, equipment, and operations meet or exceed the standards set by OSHA to avoid penalties and, more importantly, to protect workers.

In practice, adherence to OSHA standards also enhances the reputation of businesses, portraying them as responsible and reliable entities committed to the safety of their employees. This helps retain skilled workers and attracts new business opportunities where safety standards are a priority.

By making OSHA compliance a central aspect of their operational policies, electrical firms safeguard their workforce and contribute to the broader goal of creating safer, more efficient work environments across the industry.

7.5 Emergency Response: First Aid and Fire Safety Protocols

In electrical work, being prepared for emergencies is not just a regulatory requirement—it's a critical aspect of everyday safety. This preparation includes having robust first aid and fire safety protocols in place. These protocols are essential for responding effectively to accidents, ensuring that injuries can be addressed immediately and efficiently to prevent serious consequences.

First aid training is fundamental for electricians. Electricians must be proficient in administering basic first aid, given the nature of their work, which often involves risks of shocks, burns, or falls. This includes knowing how to treat burns, perform CPR, and manage shock victims until professional medical help arrives. It is also crucial to equip work sites with well-stocked first aid kits and ensure all personnel know their locations and how to use the contents.

Fire safety is another critical area. Electrical fires can result from faulty wiring, overloaded circuits, or failure to follow proper installation protocols. Thus, electricians must be trained in fire prevention techniques and know how to react if a fire occurs. This involves understanding the types of fire extinguishers and their correct use depending on the nature of the fire—whether electrical, chemical, or standard combustible materials are involved.

Regular drills and training sessions should also be conducted to keep everyone's knowledge fresh and ensure all team members react correctly under pressure. These drills can help identify potential weaknesses in emergency plans, providing a chance to refine strategies and ensure that everyone knows their role in an emergency.

Implementing comprehensive emergency response protocols minimizes the risk of severe injuries or damage during an electrical incident. It instills a sense of confidence and safety among workers, allowing them to perform their duties without fear, knowing they are well-prepared to handle unexpected situations. This proactive approach to workplace safety underscores a commitment to the well-being of every individual on the job, fostering a culture of caution and care that transcends individual actions to become a cornerstone of the workplace environment.

CHAPTER 8

ADVANCED MOTOR CONTROLS AND AUTOMATION

Advanced motor controls and automation represent a critical frontier in electrical engineering, blending sophisticated technology with practical applications to enhance the efficiency and functionality of motors across various industries. This area covers multiple topics, from fundamental control techniques to complex automation systems, equipping professionals with the necessary skills to implement these advanced technologies. As industries increasingly rely on automated processes and precise motor control for improved performance and innovation, understanding these systems becomes essential for any electrician or engineer looking to excel in the modern technological landscape.

8.1 Basics of Motor Controls

Understanding the basics of motor controls is essential for anyone installing, maintaining, or troubleshooting motor-driven systems. Motor controls are designed to start, stop, and regulate the speed and torque of electric motors, ensuring they operate efficiently and effectively across various applications.

At the core of motor control technology are starters and controllers, which manage the electrical power supplied to motors. Starters are fundamental components that safely engage and disengage the engine from the power supply, protecting it from electrical surges and helping to avoid mechanical stress. This is particularly important in industrial settings where motors are frequently turned on and off, and smooth ramp-up and ramp-down processes can significantly extend the motor's life.

Controllers further refine the functionality of motor systems by adjusting operational parameters like speed and torque. This control is critical in applications where precision is necessary, such as conveyor systems or automated production lines. Controllers can optimize performance and energy efficiency by fine-tuning how a motor operates, leading to significant cost savings and improved system reliability.

Integrating electronic components, such as programmable logic controllers (PLCs) and motor management systems, has transformed traditional motor control into a more automated and intelligent. These technologies allow for real-time monitoring and adjustments, providing a high level of control and feedback that was impossible with older systems. For instance, a PLC can be programmed to handle tasks such as motor start sequencing, speed control, and overload protection while communicating with other parts of a production system.

For electricians and technicians, mastering the basics of motor controls involves understanding the mechanical and electrical components and learning how to leverage modern electronics to enhance system performance. This knowledge is crucial for keeping up with advancements in the field and ensuring that motor-driven systems are safe and optimally configured to meet the demands of today's industries.

8.2 Different Types of Motor Starters

Motor starters are crucial components in the electrical industry. They are designed to safely activate and deactivate motors while providing necessary protection during operation. Several types of motor starters exist, each tailored to meet specific operational needs and environments, and understanding these variations is critical to selecting the right starter for any application.

Direct-on-line (DOL) starters are the simplest type of motor starters. They connect the motor directly to the power supply, providing full voltage to the system. This type of starter is widely used because of its simplicity and cost-effectiveness. However, because it initiates full torque and current immediately upon startup, it is best suited for smaller motors that do not require gradual ramp-up times.

Star-delta starters are designed for larger motors that need a softer start to minimize mechanical stress and electrical spikes. This starter operates in two stages: initially, it connects the motor windings in a star configuration to reduce voltage and subsequently switches to a delta arrangement to allow full power operation. This method reduces the initial current surge, providing a gentler start and reducing wear on the motor components.

Soft starters are advanced devices that control the voltage supplied to the motor, gradually increasing it to allow a smooth acceleration to full speed. This method significantly reduces mechanical and electrical stress, making soft starters ideal for applications involving heavy loads or those requiring precise control over the startup phase, such as large pumps or fans.

Variable Frequency Drives (VFDs) are another sophisticated type of motor starter that not only controls the starting and stopping of a motor but also adjusts its speed and torque during operation. VFDs can finely tune motor speed by varying the frequency of the power supplied to the motor, optimizing performance and energy efficiency. These starters are particularly beneficial in applications requiring variable speed operation and are instrumental in applications ranging from conveyor systems to HVAC units.

Choosing the right motor starter involves considering factors such as the motor size, the mechanical load characteristics, and the operational environment. Each type of starter offers distinct advantages, and the selection must align with both the specific requirements of the application and the overall goals of energy efficiency and motor longevity. This understanding is crucial for electricians and engineers responsible for designing and maintaining motor-driven systems.

8.3 Role and Functionality of Relays and Contactors

Relays and contactors are pivotal in electrical systems, critical components for controlling power circuits and protecting equipment. Both devices function as switches but are suited to different applications and power levels, making understanding their roles and functionalities essential for any electrician or engineer designing or maintaining electrical systems.

Relays are electrically operated switches used primarily in low to medium-power circuits. They use a small electrical current to control a much larger current, ensuring that systems can be operated safely and with minimal physical effort. For instance, a relay in a residential heating system can allow a low-voltage thermostat circuit to control a high-voltage furnace switch. This separation of circuits in the control and load sides provides both safety and functional advantages, allowing for using sensitive or low-power devices to control heavier electrical loads without direct interaction.

Conversely, contactors are designed to handle higher current loads and are commonly used in commercial and industrial applications. They function similarly to relays but are built to manage the robust demands of systems like large motors or lighting installations. Contactors often integrate additional features such as overload protection relays, enhancing their utility in protecting expensive industrial equipment from damage caused by current overload.

One of the critical functionalities of both relays and contactors is their ability to perform their switching operations remotely and automatically. This capability is crucial in systems requiring integrated controls, automated operations, and safety interlocks. For example, in industrial manufacturing lines, contactors can start multiple motors sequentially based on a programmed schedule or in response to conditions sensed by relays, such as temperature or pressure changes.

Relays and contactors thus simplify the management of electrical systems by providing efficient control over power distribution and enhancing safety by reducing the need for direct manual control of high-power circuits. Professionals can ensure more reliable, efficient, and safe electrical installations by effectively understanding and utilizing these components.

8.4 Introduction to Automation Systems

Automation systems represent a significant advancement in the way industries and even residential environments operate. They streamline processes and enhance efficiency and control. These systems integrate various technologies to automatically manage equipment and processes, reducing the need for human intervention and thereby increasing precision and productivity.

At the heart of automation systems are controllers, such as Programmable Logic Controllers (PLCs) or computer-based systems, which serve as the brains of the operation. These controllers execute pre-programmed sequences of operations at speeds and accuracy that are unachievable by human operators. For example, a PLC might control the timing and operation of conveyor belts, robotic arms, and other machinery in a manufacturing plant, coordinating exact production processes.

Sensors and actuators form the other critical components of automation systems. Sensors collect data from the environment—such as temperature, pressure, and proximity—and send it to the controllers. This data informs the controllers about the system's current conditions, enabling them to make decisions and adjust operations dynamically. Conversely, actuators are the devices that carry out actions based on the controller's commands. They can open valves, start motors, or adjust machine positions to meet the system's needs.

Connectivity and integration technologies also play vital roles, allowing different parts of an automation system to communicate seamlessly and operate as a cohesive unit. Advances in network technologies and standards, such as Industrial Ethernet and wireless communication, have expanded the capabilities of automation systems, enabling more complex and integrated operations that can be managed remotely.

The introduction of automation systems drastically changes an environment's operational dynamics. In industrial settings, automation boosts production capacity and enhances safety by removing workers from dangerous tasks. Automation can significantly improve energy efficiency and user comfort in homes and commercial buildings. As technology advances, the role of automation systems is set to grow, making their understanding and implementation increasingly crucial for electrical professionals looking to stay at the forefront of their field.

8.5 Advanced Motor Control and Efficiency

Advanced motor control and efficiency are paramount in modern electrical engineering, driven by the need for more sustainable practices and energy savings. This area of expertise involves employing sophisticated technologies and methods to optimize motors' performance and energy consumption, which are among the most significant users of electrical energy in industrial applications.

Today, motor control has evolved beyond simple start-and-stop mechanisms. For instance, variable frequency drives (VFDs) are at the forefront of advanced motor control. These devices precisely adjust the speed of an electric motor based on the demands of the load it's driving. By allowing the motor speed to vary, VFDs significantly reduce energy consumption, as the engine only uses the energy necessary for the current load rather than running at full speed continuously.

Another innovative approach is the use of soft starters. Soft starters temporarily reduce the load and torque in the powertrain and electric current surge of the motor during start-up. This smoother approach extends the motor's life by reducing mechanical stress and heat generated during the start-up phase. The result is a longer-lasting motor, lower maintenance costs, and improved energy efficiency.

The integration of intelligent sensors into motor control systems represents another leap forward. These sensors provide real-time data on motor performance and conditions, allowing for predictive maintenance and more refined control strategies. For example, sensors can detect imbalances or overheating before they lead to failures, prompting preemptive action that avoids downtime and saves energy.

These advancements in motor control technology boost operational efficiency and contribute to broader environmental goals by reducing energy use and minimizing waste. For businesses, the benefits are twofold: cost savings on energy and operations and compliance with increasingly stringent environmental regulations.

Understanding these advanced motor control technologies is essential for electricians and engineers. It enables them to implement systems that meet the dual demands of enhanced performance and sustainability, marking a significant step forward in electrical system design and management.

CHAPTER 9

DESIGN AND MANAGEMENT OF ELECTRICAL FEEDERS

Designing and managing electrical feeders is crucial for ensuring efficient, safe, and reliable power distribution across various installations. Electrical feeders are the main arteries of power distribution systems, channeling electricity from one central point to multiple circuits within a building or across a facility. This topic covers the comprehensive principles, practices, and technical considerations necessary to effectively plan, install, and maintain these systems. For professionals in electrical engineering and installation, mastering this area is essential for optimizing the performance and safety of power distribution networks.

9.1 Introduction to Electrical Feeders

Electrical feeders are fundamental components in any power distribution system. They serve as the primary pathways for electrical energy from one central location to various circuits or systems within a building or across a facility. Understanding these vital elements is essential for any professional involved in designing, installing, and maintaining electrical systems.

Feeders are larger cables or conduits that carry electricity from the main service panel or subpanel to branch circuits or other panels. They do not include branch circuits; instead, they enable power distribution to these circuits, making them a critical backbone of any electrical infrastructure. The feeders' design and capacity directly influence the electrical system's overall efficiency and safety.

Proper feeder sizing and routing are crucial to accommodating the anticipated electrical load while minimizing losses and preventing potential hazards such as overheating. This requires careful calculation and an understanding of load demands, conductor materials, and environmental factors that might affect performance. Additionally, feeders must comply with national and local electrical codes, which dictate specific requirements to ensure safety and functionality.

In practical terms, electrical feeders must be robust and reliable. They often run through concealed spaces such as conduits or are buried underground, making them less accessible for regular inspection. As a result, the initial design and installation require meticulous attention to detail, ensuring that every aspect of the feeder system is optimized for long-term reliability and safety.

For electricians and engineers, a deep understanding of electrical feeders and their management is about more than just technical competence. It's about ensuring that every part of the electrical distribution system works harmoniously to distribute power efficiently and safely, supporting the functionality of the entire electrical infrastructure.

9.2 Electrical Feeder Design and Application

Designing electrical feeders is a critical task that requires precise calculations and a thorough understanding of a facility's current and future power needs. The process involves determining the optimal path and size for the feeder cables that will carry electricity from the service panel to various subsystems and circuits, ensuring efficiency and safety.

The starting point in feeder design is assessing the total electrical load that the system will need to support. This involves considering the present requirements and anticipating potential future expansions. Calculations must account for the maximum demand that might be placed on the system under peak usage scenarios. This foresight prevents the costly and disruptive need to retrofit the system later as power demands grow.

Another crucial aspect is selecting the suitable materials and specifications for feeder cables. Factors like voltage drop, ambient temperature, and physical obstructions influence cable type, insulation, and size decisions. For instance, cables running through hot environments require robust insulation to prevent degradation and maintain performance, while those in areas prone to mechanical damage might need additional protective sheathing.

The routing of feeders also demands careful planning. The route must minimize exposure to potential hazards, such as water or corrosive chemicals, and avoid electromagnetic interference from nearby equipment. Efficient routing helps reduce installation costs and improves the safety and reliability of the electrical supply.

Compliance with national and local electrical codes is non-negotiable. These regulations dictate specifics regarding feeder installation, including clearance from other utilities, burial depth for underground cables, and the types of conduits or enclosures needed. Staying updated with these regulations ensures that the electrical feeder system is efficient and legally compliant, minimizing the risk of penalties or enforced modifications during inspections.

Electrical feeder design is a sophisticated blend of technical calculations, practical planning, and regulatory compliance. Mastery in this area allows professionals to create systems that can meet today's demands and be adaptable enough to accommodate tomorrow's growth and changes.

9.3 Sizing, Selection, and Installation of Feeders

Sizing, selecting, and installing electrical feeders are critical to ensure a power distribution system functions efficiently and safely. These processes involve careful planning and consideration of various factors that affect the electrical system's performance and reliability.

When sizing feeders, the primary consideration is the electrical load that the feeder needs to handle. This calculation is based on the total demand for all circuits the feeder will supply. Engineers use specific formulas to determine the appropriate gauge and material of the feeder cables, ensuring they can handle the maximum expected current without significant voltage drops or overheating. Factors such as the length of the feeder, the type of load (whether continuous or intermittent), and the ambient conditions also influence these calculations.

Selection of the suitable feeder cable involves more than just sizing; the material of the cable and its insulation are also crucial. For example, copper is popular due to its excellent electrical conductivity and durability. Still, it's more expensive than aluminum, which is lighter and used in applications where cost is a more significant concern. The insulation type must suit the environmental conditions, such as resistance to heat, moisture, and chemicals, ensuring the feeder's long-term reliability and safety.

The installation process must follow stringent codes and standards to ensure safety and efficiency. This includes deciding on the installation method, such as whether the feeder will be run overhead, in conduit, or buried underground. Each technique has specific requirements regarding protection, spacing, and support. For instance, underground feeders require robust insulation and protection against moisture and physical damage.

Moreover, ensuring proper grounding and bonding is essential to protect against electrical faults and enhance system stability during installation. This involves connecting the feeder system to the ground at specific points to ensure any fault current can safely dissipate into the earth, reducing the risk of electric shock or fire.

By meticulously managing these aspects—sizing, selection, and installation—professionals can guarantee that the feeder not only meets the current demands of the facility but is also capable of accommodating future expansions, ensuring a robust and adaptable power distribution network.

9.4 Overcurrent Protection and Voltage Drop Considerations

Addressing overcurrent protection and voltage drop considerations is crucial when designing and installing electrical feeder systems. These factors directly influence the electrical network's safety, efficiency, and longevity, ensuring that it operates within safe parameters while delivering optimal performance.

Overcurrent protection is essential to prevent potential damage when too much current flows through a feeder. This can happen due to overloads, short circuits, or ground faults, which pose significant risks to the electrical system and connected equipment. The primary devices used for overcurrent protection include circuit breakers and fuses, designed to interrupt the flow of electricity if the current exceeds safe levels. Selecting the right size and type of these protective devices is based on the maximum expected current load and the specific characteristics of the feeder and connected circuits. This ensures the protective device will quickly cut off the current, thus preventing potential damage or fire.

Voltage drop considerations are equally important, particularly in long feeder runs or high-load scenarios. Voltage drop occurs when the transmitted electrical energy is reduced as it travels along the length of a conductor, resulting in lower voltage levels at the endpoints than at the source. This drop can lead to inefficient operation or failure of electrical equipment that does not receive sufficient voltage. To mitigate this, calculations must be made based on the feeder's total length, the conductor's material, the cross-sectional area, and the load current. These calculations help select the appropriate wire size to minimize voltage drop without unnecessarily oversizing the wire, which can be cost-prohibitive.

Both overcurrent protection and voltage drop are critical elements that need careful consideration during the feeder design and installation. Ensuring proper overcurrent protection safeguards the electrical system from damage and hazards while effectively managing voltage drop guarantees that all system parts receive the necessary power to function correctly. Integrating these considerations seamlessly into the feeder design enhances the electrical installation's overall safety, functionality, and efficiency, providing reliable power delivery and extending the system's operational life.

9.5 Grounding, Bonding, and Maintenance of Feeders

Grounding, bonding, and regular maintenance are fundamental to managing electrical feeders and ensuring their safe and efficient operation. Each component plays a crucial role in the overall stability and performance of electrical systems, and understanding how to implement these processes properly is vital for any electrician.

Grounding involves connecting the electrical system to the earth using ground wires that extend to a grounding rod or plate. This connection provides a safe path for fault currents, helping to prevent electric shock or fire in case of a fault in the system. It's a safety measure that dissipates unwanted voltage, ensuring that the electrical system does not pose a hazard to users or equipment.

Bonding complements grounding by ensuring that all conductive parts within an electrical system, such as metal enclosures and frames, are electrically connected and have the same electrical potential. This prevents voltage differences between conductive parts that can lead to shock hazards. Adequate bonding is critical in environments where various components of an electrical system are separated by distance but need to work as a cohesive unit, ensuring safety across the entire network.

Maintenance of feeders is essential to preserve their integrity and functionality over time. Regular inspections are necessary to identify and address issues such as corrosion, damage to insulation, or loose connections that can compromise safety and efficiency. Such checks help in the early detection of potential problems, preventing costly repairs or dangerous situations down the line. Maintenance routines should also include testing for electrical continuity and resistance to ensure that grounding and bonding connections remain effective even after years of service.

For electricians, implementing thorough grounding and bonding practices, combined with a proactive maintenance schedule, ensures that feeder systems are compliant with safety standards and capable of delivering reliable performance. This proactive approach minimizes downtime, enhances system safety, and extends the lifespan of the electrical infrastructure, providing peace of mind and significant cost savings over time.

PART 2

PRACTICE

Part 2: Practice focuses on applying the theoretical knowledge gained in the earlier sections to real-world scenarios, offering a hands-on approach to mastering electrical tasks. This section provides practical exercises, detailed troubleshooting scenarios, and step-by-step guides, equipping readers with the essential skills needed to proficiently handle both everyday tasks and complex challenges in the field of electrical engineering. It's designed to bridge the gap between understanding and applying concepts effectively, ensuring that professionals are well-prepared to implement their knowledge practically and efficiently.

Journeyman Electrician Exam PRACTICE (500+ Exam Questions)

The "Journeyman Electrician Exam Practice" section provides an extensive collection of over 500 exam questions meticulously designed to ensure comprehensive preparation for the Journeyman Electrician Exam. Here, candidates will find a wide range of question types that cover the full spectrum of topics required for certification. This practice set is essential for reinforcing the concepts learned throughout the course, honing problem-solving skills, and boosting test-taking confidence. It is an invaluable tool for anyone aiming to succeed on the exam and excel in their electrical career.

Specific Questions for each topic

The practice section includes specific questions for each topic, designed to thoroughly test your knowledge in distinct areas crucial for the Journeyman Electrician Exam. This setup allows you to engage deeply with each subject, from electrical theory and code requirements to safety practices and practical applications. By tackling these focused questions, you can pinpoint areas where you might need more review and reinforce your understanding where it's strongest. This targeted approach ensures you're well-prepared across all necessary exam topics, enhancing your confidence and competence.

National Electrical Code (NEC) Questions

Q1: What is the minimum required burial depth for non-metallic sheathed cable in a residential front yard without a raceway?

A. 6 inches
B. 12 inches
C. 18 inches
D. 24 inches

Q2: According to the NEC, what is the maximum allowable voltage drop for a branch circuit?

A. 2%
B. 3%
C. 5%
D. 8%

Q3: Which NEC article addresses the installation and requirements for grounding electrodes?

A. Article 200
B. Article 250
C. Article 300
D. Article 450

Q4: In NEC terminology, what is considered a 'continuous load'?

A. Any load expected to last for less than 15 minutes
B. Any load expected to last for more than 3 hours
C. Any load that varies in power consumption
D. Any load operating at maximum capacity periodically

Q5: Where must GFCI protection be provided in residential dwellings according to the NEC?

A. Only in bathrooms
B. In bathrooms and kitchens
C. In all wet or damp locations
D. Only outdoors

Q6: What is the minimum wire size generally allowed for residential lighting circuits according to the NEC?

A. 18 AWG
B. 16 AWG
C. 14 AWG
D. 12 AWG

Q7: NEC requires arc-fault circuit interrupter (AFCI) protection in which location(s)?

A. Bedrooms only
B. Kitchen and bathrooms
C. All living areas except bathrooms
D. Garages and outdoors

Q8: What is the minimum height for installing outlets in a habitable room according to the NEC?

A. 12 inches from the floor
B. 15 inches from the floor
C. 18 inches from the floor
D. No specific requirement

Q9: According to the NEC, what is the required clearance in front of electrical panels?

A. 30 inches
B. 36 inches
C. 42 inches
D. 48 inches

Q10: When are conductors considered to be protected from physical damage according to the NEC?

A. When installed within a wall
B. When enclosed in conduit
C. When covered by drywall
D. When insulated with non-metallic sheathing

Q11: According to the NEC, under which condition can NM cable be exposed without conduit protection?

A. In commercial garages
B. In residential basements
C. In damp locations
D. It cannot be exposed without protection in any condition

Q12: Which NEC article covers the requirements for branch circuits?

A. Article 200
B. Article 210
C. Article 215
D. Article 220

Q13: What does the NEC require for the identification of the grounded conductor?

A. Must be colored green
B. Must be colored white or gray
C. Must be labeled with tape
D. No specific requirement

Q14: According to the NEC, which color is designated for the ground conductor?

A. Green
B. White
C. Black
D. Red

Q15: What maximum current is allowed for a 14 AWG copper wire according to the NEC?

A. 15 amps
B. 20 amps
C. 25 amps
D. 30 amps

Q16: How frequently must flexible cords be supported when used in a temporary installation according to the NEC?

A. Every 4 feet
B. Every 6 feet
C. Every 10 feet
D. Every 12 feet

Q17: According to the NEC, which appliance must have a dedicated circuit?

A. Dishwasher
B. Refrigerator
C. Microwave
D. All of the above

Q18: What is the minimum required overhead clearance of service conductors over a residential driveway according to the NEC?

A. 10 feet
B. 12 feet
C. 14 feet
D. 18 feet

Q19: According to the NEC, what is the minimum required receptacle spacing in a residential kitchen countertop?

A. Every 6 feet
B. Every 4 feet
C. Every 2 feet
D. No minimum required if all appliances are fixed

Q20: What type of protection is required for all 15A and 20A, 120V receptacles installed in a children's playroom according to the NEC?

A. GFCI
B. AFCI
C. Both GFCI and AFCI
D. No specific protection is required

Q21: In the NEC, what is the maximum allowable gap between enclosures and barriers for fire-rated assemblies?

A. 1/8 inch
B. 1/4 inch
C. 1/2 inch
D. 3/4 inch

Q22: What NEC requirement applies to the installation of cables in vertical shafts?

A. Must use armored cable
B. Must use plenum-rated cable
C. Must be enclosed in conduit
D. No specific requirement

Q23: According to the NEC, when is the use of aluminum wiring permitted for branch circuit wiring in residential applications?

A. Never allowed
B. Always allowed
C. Allowed for 240V circuits only
D. Allowed if the wire gauge is 8 AWG or larger

Q24: Which NEC rule applies to the identification of switch loops?

A. The hot wire must be black
B. The neutral wire must be white
C. The ground wire must be green
D. The hot wire must be red

Q25: According to the NEC, how high must the disconnecting means be installed for appliances?

A. Within sight from the appliance location
B. Not more than 6 feet above the floor
C. Not more than 6 feet 7 inches above the floor
D. No height requirement as long as it is accessible

Q26: What is the NEC requirement for the minimum number of lighting outlets in a residential hallway?

A. One per every 10 feet
B. One at each end
C. One every 50 feet
D. No specific requirement if illuminated by adjacent rooms

Q27: What is the minimum size copper conductor typically used for grounding in residential wiring?

A. 8 AWG
B. 10 AWG
C. 12 AWG
D. 14 AWG

Q28: In the NEC, what is the required minimum clearance from the top of a cooking range to a non-combustible range hood?

A. 24 inches
B. 30 inches
C. 36 inches
D. 42 inches

Q29: Which NEC article addresses the requirements for bathrooms in residential buildings?

A. Article 210
B. Article 215
C. Article 220
D. Article 240

Q30: How often must emergency and standby power systems be tested according to the NEC?

A. Weekly
B. Monthly
C. Quarterly
D. Annually

Q31: According to the NEC, which type of conduit is NOT permitted for underground installations?

A. Rigid Metal Conduit (RMC)
B. Electrical Nonmetallic Tubing (ENT)
C. Intermediate Metal Conduit (IMC)
D. Flexible Metal Conduit (FMC)

Q32: What is the maximum length permitted for a flexible cord used with a portable appliance?

A. 3 feet
B. 6 feet
C. 12 feet
D. No specific limit as long as it is deemed safe

Q33: NEC requires how many watts per square foot for general lighting in commercial buildings?

A. 1 watt per square foot
B. 2 watts per square foot
C. 3 watts per square foot
D. 5 watts per square foot

Q34: According to the NEC, what minimum wire size is required for a 20-amp circuit?

A. 12 AWG
B. 14 AWG
C. 16 AWG
D. 18 AWG

Q35: What NEC article covers the installation and requirements for communication circuits?

A. Article 760
B. Article 770
C. Article 800
D. Article 810

Q36: What is the minimum allowable size for service entrance conductors in residential applications according to the NEC?

A. 8 AWG
B. 10 AWG
C. 12 AWG
D. 14 AWG

Q37: Which NEC requirement must be met for the installation of overcurrent protection devices in a dwelling?

A. Must be readily accessible
B. Can be installed behind movable panels
C. Can be located in bathrooms
D. All of the above

Q38: How many conductors are allowed under a single terminal of a device according to the NEC?

A. One
B. Two
C. Three
D. Depends on the manufacturer's rating

Q39: According to the NEC, what is the minimum required insulation rating for wiring in residential building walls?

A. THHN
B. THWN
C. XHHW
D. USE

Q40: What does the NEC require for all receptacles installed in children's rooms?

A. Tamper-resistant receptacles
B. Weather-resistant receptacles
C. AFCI protection
D. GFCI protection

Q41: According to the NEC, what is the maximum height for switch installation in public buildings?

A. 40 inches
B. 48 inches
C. 54 inches
D. 60 inches

Q42: What NEC article deals with the safety requirements for swimming pool installations?

A. Article 680
B. Article 690
C. Article 700
D. Article 710

Q43: In the NEC, what is the recommended practice for identifying a high-leg of a delta system in panelboards?

A. Marked with blue tape
B. Marked with orange tape
C. Marked with red tape
D. No specific color required

Q44: According to the NEC, what is the minimum required rating for a circuit breaker in a typical residential home?

A. 15 amps
B. 20 amps
C. 30 amps
D. Depends on the specific appliances and total load

Q45: NEC specifies what type of protection for circuits supplying lighting fixtures in closets?

A. Arc-fault circuit interrupter (AFCI)
B. Ground-fault circuit interrupter (GFCI)
C. Surge protection
D. Overcurrent protection

Q46: According to the NEC, what is the minimum clearance required around an electric meter socket?

A. 3 feet
B. 6 feet
C. 12 feet
D. No specific requirement

Q47: What NEC article provides guidelines for the installation of generators?

A. Article 445
B. Article 450
C. Article 455
D. Article 460

Q48: According to the NEC, how far must receptacles be from the edge of a bathtub or shower?

A. At least 3 feet
B. At least 6 feet
C. At least 8 feet
D. Not allowed within the same room

Q49: What is the minimum wire bending space required by the NEC at terminals?

A. Three times the wire diameter
B. Six times the wire diameter
C. Eight times the wire diameter
D. No specific requirement, depends on the terminal

Q50: According to the NEC, under what condition is the use of non-metallic sheathed cable permitted in commercial buildings?

A. Above suspended ceilings
B. In all types of walls
C. In plenum spaces
D. Only in non-plenum return air ceilings

Q51: What NEC rule applies to the color identification of neutral conductors?

A. Must be colored blue
B. Must be colored white or gray
C. Must be colored green
D. No specific color as long as it is distinguishable

Q52: According to the NEC, what is the required minimum number of circuits for lighting and receptacles in a residential kitchen?

A. One
B. Two
C. Three
D. Four

Answers

Correct Answer 1: C. 18 inches

Explanation: The NEC requires a minimum burial depth of 18 inches for non-metallic sheathed cable without a raceway in residential front yards to protect the cable from physical damage.

Correct Answer 2: B. 3%

Explanation: The NEC recommends a maximum voltage drop of 3% for branch circuits to ensure efficient operation of electrical devices without significant loss of performance.

Correct Answer 3: B. Article 250

Explanation: NEC Article 250 covers all aspects related to grounding and bonding, providing detailed requirements for grounding electrodes, methods, and components.

Correct Answer 4: B. Any load expected to last for more than 3 hours

Explanation: The NEC defines a continuous load as any load expected to continue for three hours or more, requiring special consideration in circuit design and capacity calculations.

Correct Answer 5: C. In all wet or damp locations

Explanation: The NEC requires GFCI protection in all areas susceptible to moisture, including bathrooms, kitchens, and outdoor spaces, to prevent electrical shock.

Correct Answer 6: C. 14 AWG

Explanation: The minimum wire size allowed for residential lighting circuits is typically 14 AWG, according to the NEC, ensuring safety and adequacy in handling typical household lighting loads.

Correct Answer 7: C. All living areas except bathrooms

Explanation: NEC requires AFCI protection in all living areas of new constructions to prevent electrical fires caused by arc faults, with the exception of bathrooms.

Correct Answer 8: B. 15 inches from the floor

Explanation: The NEC typically requires outlets to be installed at least 15 inches from the floor to be accessible yet safe from common household hazards.

Correct Answer 9: B. 36 inches

Explanation: The NEC requires a minimum of 36 inches of clearance in front of electrical panels to ensure safety during maintenance and operation, allowing enough space for clear access and egress.

Correct Answer 10: B. When enclosed in conduit

Explanation: Conductors are considered protected from physical damage when enclosed in a conduit, which offers robust protection against impacts and environmental elements.

Correct Answer 11: B. In residential basements

Explanation: NM cable can be exposed without additional protection in residential basements where it is generally safe from physical damage and moisture, according to the NEC.

Correct Answer 12: B. Article 210

Explanation: NEC Article 210 specifically addresses the requirements for branch circuits, including their ratings, load calculations, and protective measures.

Correct Answer 13: B. Must be colored white or gray

Explanation: The NEC requires the grounded (neutral) conductor to be identified with white or gray color to distinguish it from ungrounded (hot) and grounding conductors.

Correct Answer 14: A. Green

Explanation: The NEC designates green (or bare wire without insulation) as the color for grounding conductors to ensure they are easily recognizable.

Correct Answer 15: A. 15 amps

Explanation: According to the NEC, a 14 AWG copper wire is generally rated for a

70

maximum current of 15 amps, suitable for most light to medium load circuits in residential wiring.

Correct Answer 16: C. Every 10 feet

Explanation: The NEC requires that flexible cords be supported every 10 feet when used in a temporary installation to prevent sagging and reduce the risk of damage or disconnection.

Correct Answer 17: D. All of the above

Explanation: The NEC mandates that certain appliances such as dishwashers, refrigerators, and microwaves each have a dedicated circuit to prevent overloading and ensure safe operation.

Correct Answer 18: B. 12 feet

Explanation: Service conductors running over a residential driveway must have a minimum overhead clearance of 12 feet according to the NEC to ensure safety and prevent accidental contact.

Correct Answer 19: B. Every 4 feet

Explanation: The NEC requires receptacles to be spaced no more than 4 feet apart along kitchen countertops to provide convenient access to power for appliances without the excessive use of extension cords.

Correct Answer 20: C. Both GFCI and AFCI

Explanation: For enhanced safety, the NEC requires both GFCI and AFCI protection for receptacles installed in children's playrooms to protect against both ground faults and arc faults.

Correct Answer 21: A. 1/8 inch

Explanation: The NEC specifies a maximum allowable gap of 1/8 inch between enclosures and barriers in fire-rated assemblies to maintain the integrity of the fire barrier.

Correct Answer 22: C. Must be enclosed in conduit

Explanation: Cables installed in vertical shafts must be enclosed in conduit according to the NEC to protect against fire spread and physical damage within building risers.

Correct Answer 23: D. Allowed if the wire gauge is 8 AWG or larger

Explanation: The NEC permits the use of aluminum wiring for branch circuit wiring in residential applications if the wire gauge is 8 AWG or larger due to its sufficient current-carrying capacity and reduced risk of overheating.

Correct Answer 24: D. The hot wire must be red

Explanation: In switch loops, the NEC requires that the hot wire be identified with red insulation to clearly indicate that it is carrying power from the source.

Correct Answer 25: B. Not more than 6 feet above the floor

Explanation: Disconnecting means for appliances must be installed not more than 6 feet above the floor to ensure they are readily accessible for operation without the need for tools or ladders.

Correct Answer 26: B. One at each end

Explanation: The NEC requires a minimum of one lighting outlet at each end of a hallway to ensure adequate illumination for safety and convenience in residential settings.

Correct Answer 27: B. 10 AWG

Explanation: 10 AWG copper conductor is commonly used for grounding in residential wiring to ensure safety and compliance with electrical standards.

Correct Answer 28: B. 30 inches

Explanation: The NEC requires a minimum clearance of 30 inches from the top of a cooking range to a non-combustible range hood to prevent fire hazards and ensure proper ventilation.

Correct Answer 29: C. Article 220

Explanation: NEC Article 220 addresses the specific requirements for bathrooms in residential buildings, focusing on load calculations and safety provisions.

Correct Answer 30: B. Monthly

Explanation: Emergency and standby power systems must be tested monthly according to the NEC to ensure they are functional and reliable in case of an emergency.

Correct Answer 31: D. Flexible Metal Conduit (FMC)

Explanation: FMC is not permitted for underground installations as it may not provide adequate protection against moisture and physical damage.

Correct Answer 32: D. No specific limit as long as it is deemed safe

Explanation: The NEC does not specify a maximum length for flexible cords used with portable appliances, as long as the setup is safe and meets other applicable requirements.

Correct Answer 33: A. 1 watt per square foot

Explanation: The NEC requires at least 1 watt per square foot for general lighting in commercial buildings to ensure adequate illumination.

Correct Answer 34: A. 12 AWG

Explanation: 12 AWG is the minimum wire size required for a 20-amp circuit to safely handle the electrical load without overheating.

Correct Answer 35: C. Article 800

Explanation: NEC Article 800 covers the installation and requirements for communication circuits, including their protection and proper routing.

Correct Answer 36: B. 10 AWG

Explanation: 10 AWG is the minimum allowable size for service entrance conductors in residential applications, ensuring adequate capacity for household electrical loads.

Correct Answer 37: A. Must be readily accessible

Explanation: Overcurrent protection devices must be readily accessible to allow for quick and safe operation, testing, and maintenance.

Correct Answer 38: A. One

Explanation: The NEC generally allows only one conductor under a single terminal to prevent loose connections and ensure secure and reliable contact.

Correct Answer 39: B. THWN

Explanation: THWN insulation is suitable for residential building walls, offering protection against moisture and heat, which are common in such environments.

Correct Answer 40: A. Tamper-resistant receptacles

Explanation: Tamper-resistant receptacles are required in children's rooms to prevent accidental shocks and injuries, enhancing safety in environments frequented by children.

Correct Answer 41: B. 48 inches

Explanation: The maximum height for switch installation in public buildings is 48 inches, making switches accessible while complying with ADA guidelines.

Correct Answer 42: A. Article 680

Explanation: NEC Article 680 details the safety requirements for swimming pool installations, addressing aspects like wiring, equipment, and bonding to enhance safety around water.

Correct Answer 43: B. Marked with orange tape

Explanation: The high-leg of a delta system must be marked with orange tape in panelboards to warn and prevent misconnections and hazards.

Correct Answer 44: A. 15 amps

Explanation: A minimum of 15 amps is typically required for circuit breakers in residential homes to handle common electrical loads safely.

Correct Answer 45: A. Arc-fault circuit interrupter (AFCI)

Explanation: AFCI protection is specified for circuits supplying lighting fixtures in closets to prevent electrical fires caused by arc faults.

Correct Answer 46: A. 3 feet

Explanation: A minimum clearance of 3 feet is required around an electric meter socket to ensure safety and accessibility for maintenance or emergency services.

Correct Answer 47: A. Article 445

Explanation: NEC Article 445 provides guidelines for the safe installation and operation of generators, including aspects related to wiring and clearances.

Correct Answer 48: B. At least 6 feet

Explanation: Receptacles must be located at least 6 feet away from the edge of a bathtub or shower to reduce the risk of electrical shock.

Correct Answer 49: C. Eight times the wire diameter

Explanation: NEC requires a minimum wire bending space at terminals of eight times the wire diameter to prevent damage and ensure safe connections.

Correct Answer 50: D. Only in non-plenum return air ceilings

Explanation: Non-metallic sheathed cable is permitted in commercial buildings above suspended ceilings or in spaces that are not used for environmental air.

Correct Answer 51: B. Must be colored white or gray

Explanation: Neutral conductors must be identified by white or gray coloring to distinguish them from other conductors, ensuring clarity and safety in wiring practices.

Correct Answer 52: C. Three

Explanation: At least three separate circuits are required for lighting and receptacles in a residential kitchen, accommodating multiple appliances and work areas effectively.

Electrical theory

Q53: What is the unit of electrical resistance?

A. Ohm
B. Ampere
C. Volt
D. Watt

Q54: What law states that the current through a conductor between two points is directly proportional to the voltage across the two points?

A. Kirchhoff's Voltage Law
B. Ohm's Law
C. Faraday's Law
D. Joule's Law

Q55: Which component is used to store electrical energy in an electric field?

A. Resistor
B. Capacitor
C. Inductor
D. Diode

Q56: What is the formula for calculating electrical power in a circuit?

A. $P = VI$
B. $P = IR$
C. $P = V/R$
D. $P = I^2R$

Q57: Which law states that the total voltage around any closed loop must be zero?

A. Ohm's Law
B. Kirchhoff's Current Law
C. Kirchhoff's Voltage Law
D. Newton's Third Law

Q58: In an AC circuit, what is the term used to describe the opposition to current flow that includes both resistance and reactance?

A. Capacitance
B. Conductance
C. Impedance
D. Inductance

Q59: What type of circuit connection results in voltages being the same across all components?

A. Series
B. Parallel
C. Mesh
D. Ring

Q60: What is the phase difference between the voltage and current in a purely resistive AC circuit?

A. 0 degrees
B. 45 degrees
C. 90 degrees
D. 180 degrees

Q61: Which theorem allows the simplification of a linear bilateral network into a single voltage source and series resistance?

A. Thevenin's Theorem
B. Norton's Theorem
C. Superposition Theorem
D. Maxwell's Theorem

Q62: What does the term 'reactance' refer to in an AC circuit?

A. The resistance to DC currents
B. The opposition to changes in voltage
C. The opposition to changes in current
D. None of the above

Q63: What type of material is best suited for making permanent magnets?

A. Copper
B. Aluminum
C. Iron
D. Silicon

Q64: What device converts mechanical energy into electrical energy?

A. Motor
B. Generator
C. Transformer
D. Capacitor

Q65: What principle is the operation of a transformer based on?

A. Electrostatic induction
B. Electromagnetic induction
C. Thermal induction
D. Mechanical induction

Q66: Which type of motor is commonly used in household appliances?

A. Synchronous motor
B. Asynchronous motor
C. DC motor
D. Stepper motor

Q67: What does the term 'duty cycle' refer to in the context of electrical devices?

A. The ratio of active time to inactive time
B. The maximum load a device can handle
C. The total lifespan of a device
D. The efficiency of energy conversion

Q68: In a series circuit, how does the total resistance compare to the individual resistances?

A. It is equal to the sum of the individual resistances
B. It is less than the smallest individual resistance
C. It is more than the largest individual resistance
D. It is the average of the individual resistances

Q69: What is the effect of increasing the frequency on the capacitive reactance in a circuit?

A. Increases
B. Decreases
C. Remains the same
D. Becomes zero

Q70: What component is typically used to filter out high-frequency noise in electronic circuits?

A. Resistor
B. Capacitor
C. Inductor
D. Transformer

Q71: What is the primary advantage of using alternating current (AC) over direct current (DC) for power transmission?

A. Lower costs
B. Higher energy efficiency
C. Easier to convert voltages
D. Safer to handle

Q72: Which device is used to measure electrical current?

A. Ammeter
B. Voltmeter
C. Ohmmeter
D. Wattmeter

Q73: What happens to the resistance of a conductor as its temperature increases?

A. Increases
B. Decreases
C. Remains the same
D. Becomes zero

Q74: What type of circuit protection device operates by melting a wire when too much current flows through it?

A. Fuse
B. Circuit breaker
C. Relay
D. Switch

Q75: What phenomenon causes the delay in the current behind the voltage in an inductive AC circuit?

A. Capacitance
B. Inductance
C. Resistance
D. Conductance

Q76: What is a common use of a diode in an electrical circuit?

A. To increase current flow
B. To decrease voltage
C. To allow current to flow in one direction only
D. To balance resistance

Q77: What effect does adding more batteries in series have on the total voltage of the circuit?

A. Increases
B. Decreases
C. Remains the same
D. Becomes zero

Q78: How does a capacitor behave in a DC circuit after it is fully charged?

A. It blocks current flow
B. It allows continuous current flow
C. It reverses current flow
D. It dissipates the current as heat

Q79: What formula represents the power factor in an AC circuit?

A. PF = True Power / Apparent Power
B. PF = Apparent Power / True Power
C. PF = Voltage x Current
D. PF = Resistance x Current

Q80: What unit is used to measure electrical capacitance?

A. Ohm
B. Farad
C. Henry
D. Tesla

Q81: What is the primary function of a rectifier?

A. To convert AC to DC
B. To convert DC to AC
C. To reduce voltage spikes
D. To increase current flow

Q82: In what type of circuit is Kirchhoff's Current Law (KCL) primarily used?

A. Series circuits only
B. Parallel circuits only
C. Any circuit with junctions
D. Circuits without junctions

Q83: What does RMS stand for and what does it signify in electrical terms?

A. Root Mean Square, representing the effective value of AC voltage or current
B. Relative Maximum Surge, indicating the maximum current in a circuit
C. Real Maximum Speed, related to the speed of electron flow
D. Resistance Measurement Standard, a unit for measuring resistance

Q84: What component is used primarily to oppose changes in current in an AC circuit?

A. Resistor
B. Capacitor
C. Inductor
D. Transformer

Q85: Which principle explains the operation of electric motors and generators?

A. Coulomb's Law
B. Lenz's Law
C. Ohm's Law
D. Faraday's Law of Electromagnetic Induction

Q86: What type of switch is used to control a circuit from two different locations?

A. Single-pole switch
B. Double-pole switch
C. Three-way switch
D. Four-way switch

Q87: What effect does inductance have on an AC circuit?

A. It increases resistance
B. It reduces current flow
C. It causes the current to lag behind the voltage
D. It causes the voltage to lag behind the current

Q88: What is typically the main advantage of using a three-phase power supply over a single-phase power supply?

A. It is more cost-effective

B. It provides higher power density
C. It reduces electrical interference
D. It offers a constant power transfer

Q89: What is the typical use of a Zener diode in a circuit?

A. Voltage regulation
B. Current amplification
C. Signal modulation
D. Power inversion

Q90: How does a thermistor's resistance change with temperature?

A. Increases with increasing temperature
B. Decreases with increasing temperature
C. Does not change with temperature
D. Randomly changes with temperature

Q91: What is the primary purpose of a surge protector in an electrical circuit?

A. To increase voltage
B. To regulate current
C. To protect devices from voltage spikes
D. To decrease resistance

Q92: Which type of material typically makes the best electrical insulator?

A. Copper
B. Aluminum
C. Rubber
D. Iron

Q93: What is the significance of the 'skin effect' in electrical conductors?

A. It is the tendency for current to flow only on the surface of the conductor at high frequencies
B. It is the effect of insulation wearing thin over time
C. It refers to the discoloration of a conductor when overheated
D. It describes the resistance increase in a conductor over time

Q94: How do you calculate the total capacitance of capacitors in series?

A. Add the values of all capacitors
B. Use the reciprocal of the sum of the reciprocals of all capacitances
C. Multiply the values of all capacitors
D. None of the above; capacitance in series does not change

Q95: What phenomenon causes flickering lights when large appliances are turned on?

A. High resistance in the circuit
B. Voltage drops due to sudden high current draw
C. Capacitive coupling in the circuit
D. Inductive kickback from the appliance

Q96: What is the standard domestic voltage supply in the United States?

A. 110V
B. 120V
C. 220V
D. 240V

Q97: What property of a conductor decreases its resistance?

A. Increasing length
B. Decreasing cross-sectional area
C. Increasing temperature
D. Decreasing temperature

Q98: What is the function of a relay in an electrical circuit?

A. To store electric charge
B. To control a large current with a smaller current
C. To convert AC to DC
D. To regulate voltage

Q99: What safety device is used to interrupt the power when detecting an imbalance between incoming and outgoing current?

A. Fuse
B. Circuit breaker
C. Ground Fault Circuit Interrupter (GFCI)
D. Residual Current Device (RCD)

Q100: In which type of wiring is the neutral wire specifically required to be insulated?

A. High-voltage transmission
B. Low-voltage applications
C. Neutral-grounded systems
D. None of the above

Q101: What causes a phase-to-phase fault in a power system?

A. Insulation breakdown
B. Excessive load
C. Improper grounding
D. All of the above

Q102: What type of battery is commonly used in automotive applications?

81

A. Nickel-Cadmium
B. Lithium-Ion
C. Lead-Acid
D. Nickel-Metal Hydride

Q103: What is the main purpose of using a ground wire in electrical installations?

A. To carry excess current back to the panel
B. To provide a path for fault current back to the earth
C. To increase the circuit's efficiency
D. To regulate the voltage in the circuit

Q104: How does adding capacitors in parallel affect the total capacitance?

A. It decreases the total capacitance
B. It increases the total capacitance
C. It does not change the capacitance
D. It depends on the voltage

1. Answers

Here are the correct answers and explanations for the questions from Q53 to Q78 focused on Electrical Theory:

Correct Answer 53: A. Ohm

Explanation: Ohm is the unit of measurement for electrical resistance, indicating how much a material opposes the flow of electric current.

Correct Answer 54: B. Ohm's Law

Explanation: Ohm's Law states that the current through a conductor between two points is directly proportional to the voltage across the two points and inversely proportional to the resistance.

Correct Answer 55: B. Capacitor

Explanation: A capacitor is used to store electrical energy in an electric field, commonly used in various electronic circuits for filtering, buffering, and energy storage.

Correct Answer 56: A. P = VI

Explanation: The formula P = VI (where P is power, V is voltage, and I is current) is used to calculate electrical power, indicating how much work can be done by the circuit per unit time.

Correct Answer 57: C. Kirchhoff's Voltage Law

Explanation: Kirchhoff's Voltage Law states that the total sum of voltages around any closed loop in a circuit must be zero, ensuring energy conservation within the loop.

Correct Answer 58: C. Impedance

Explanation: Impedance is the term used in AC circuits to describe the total opposition to current flow, which includes both resistance (real part) and reactance (imaginary part).

Correct Answer 59: B. Parallel

Explanation: In a parallel circuit connection, voltages across all components are the same, as each component is directly connected across the power supply.

Correct Answer 60: A. 0 degrees

Explanation: In a purely resistive AC circuit, the phase difference between the voltage and current is 0 degrees, meaning they increase and decrease in sync.

Correct Answer 61: A. Thevenin's Theorem

Explanation: Thevenin's Theorem allows for the simplification of a linear bilateral network into a single voltage source and a series resistance, facilitating easier analysis and problem-solving.

Correct Answer 62: C. The opposition to changes in current

Explanation: Reactance in an AC circuit refers to the opposition that inductors and capacitors provide to changes in current due to their ability to store energy in magnetic and electric fields, respectively.

Correct Answer 63: C. Iron

Explanation: Iron is best suited for making permanent magnets due to its ferromagnetic properties, which allow it to retain a significant amount of magnetism after being magnetized.

Correct Answer 64: B. Generator

Explanation: A generator is a device that converts mechanical energy into electrical energy by moving a conductor through a magnetic field, inducing a voltage across the conductor.

Correct Answer 65: B. Electromagnetic induction

Explanation: The operation of a transformer is based on electromagnetic induction, where a changing magnetic field in the primary coil induces a voltage in the secondary coil.

Correct Answer 66: B. Asynchronous motor

Explanation: Asynchronous motors, or induction motors, are commonly used in household appliances due to their simple design, robustness, and cost-effectiveness.

Correct Answer 67: A. The ratio of active time to inactive time

Explanation: The term 'duty cycle' refers to the ratio of active time to inactive time, describing how long a device operates compared to how long it rests within a given period.

Correct Answer 68: A. It is equal to the sum of the individual resistances

Explanation: In a series circuit, the total resistance is the sum of all individual resistances, as the current must pass through each resistor sequentially.

Correct Answer 69: B. Decreases

Explanation: Increasing the frequency decreases the capacitive reactance in a circuit because capacitive reactance is inversely proportional to the frequency of the current.

Correct Answer 70: B. Capacitor

Explanation: Capacitors are typically used to filter out high-frequency noise in electronic circuits by smoothing out voltage fluctuations and providing a path to ground for the noise.

Correct Answer 71: C. Easier to convert voltages

Explanation: The primary advantage of using alternating current (AC) over direct current (DC) for power transmission is the ease of voltage conversion using transformers, which can efficiently step up or step down voltage levels.

Correct Answer 72: A. Ammeter

Explanation: An ammeter is a device used to measure electrical current, typically connected in series with a circuit to directly measure the flow of current.

Correct Answer 73: A. Increases

Explanation: The resistance of a conductor generally increases as its temperature increases, due to the increased vibration of atoms within the conductor, which impedes the flow of electrons.

Correct Answer 74: A. Fuse

Explanation: A fuse is a type of circuit protection device that operates by melting a thin wire inside it when too much current flows through, thus breaking the circuit to prevent damage.

Correct Answer 75: B. Inductance

Explanation: In an inductive AC circuit, the phenomenon of inductance causes the current to lag behind the voltage, due to the energy being temporarily stored in the magnetic field of the inductor.

Correct Answer 76: C. To allow current to flow in one direction only

Explanation: Diodes are used in electrical circuits to allow current to flow in only one direction, effectively preventing current from flowing in the reverse direction.

Correct Answer 77: A. Increases

Explanation: Adding more batteries in series increases the total voltage of the circuit because the voltages of individual batteries are summed.

Answer 78: A. It blocks current flow

Explanation: Once a capacitor is fully charged in a DC circuit, it behaves like an open circuit and blocks any further current flow.

Here are the correct answers and explanations for questions Q79 to Q104 focused on Electrical Theory:

Correct Answer 79: A. PF = True Power / Apparent Power

Explanation: Power factor in an AC circuit is defined as the ratio of true power (measured in watts) to apparent power (measured in volt-amperes), representing the efficiency with which the power is being used.

Correct Answer 80: B. Farad

Explanation: The unit of measurement for electrical capacitance is the Farad, which quantifies the amount of electric charge a capacitor can store per unit of voltage across its plates.

Correct Answer 81: A. To convert AC to DC

Explanation: A rectifier is a device used primarily to convert alternating current (AC) into direct current (DC), which is useful for powering DC electronics and charging batteries.

Correct Answer 82: C. Any circuit with junctions

Explanation: Kirchhoff's Current Law (KCL) is applied to any circuit with junctions or nodes, stating that the total current entering a junction must equal the total current leaving the junction.

Correct Answer 83: A. Root Mean Square, representing the effective value of AC voltage or current

Explanation: RMS stands for Root Mean Square and is used to represent the effective or equivalent DC value of an AC voltage or current, reflecting the actual power delivered by the AC cycle.

Correct Answer 84: C. Inductor

Explanation: An inductor is used in AC circuits primarily to oppose changes in current, storing energy in a magnetic field and releasing it, thus smoothing out changes in current flow.

Correct Answer 85: D. Faraday's Law of Electromagnetic Induction

Explanation: The operation of both electric motors and generators is based on Faraday's Law of Electromagnetic Induction, which describes how a voltage can be induced by changing the magnetic environment of a conductor.

Correct Answer 86: C. Three-way switch

Explanation: A three-way switch is designed to control a single circuit from two different locations, typically used in hallways or rooms with multiple entry points.

Correct Answer 87: C. It causes the current to lag behind the voltage

Explanation: Inductance in an AC circuit causes the current to lag behind the voltage due to the energy being temporarily stored in the magnetic field of the inductor.

Correct Answer 88: D. It offers a constant power transfer

Explanation: A three-phase power supply provides a constant power transfer which is more efficient for large-scale power distribution and heavy machinery operation, as it ensures a continuous power flow and better balance across the phases.

Correct Answer 89: A. Voltage regulation

Explanation: A Zener diode is typically used for voltage regulation in circuits, as it can maintain a steady voltage across itself when reversed biased, preventing excess voltage from damaging circuit components.

Correct Answer 90: B. Decreases with increasing temperature

Explanation: A thermistor's resistance decreases as temperature increases if it is a Negative Temperature Coefficient (NTC) thermistor, commonly used for temperature sensing and protection applications.

Correct Answer 91: C. To protect devices from voltage spikes

Explanation: A surge protector is designed to protect electrical devices from voltage spikes by diverting excess voltage away from the devices, typically through a grounding wire.

Correct Answer 92: C. Rubber

Explanation: Rubber is an excellent electrical insulator due to its high resistivity and inability to conduct electrical current, commonly used to coat wires and other conductive surfaces.

Correct Answer 93: A. It is the tendency for current to flow only on the surface of the conductor at high frequencies

Explanation: The 'skin effect' is significant at high frequencies where the current tends to flow only at the surface of the conductor, reducing the effective cross-sectional area available for current flow.

Correct Answer 94: B. Use the reciprocal of the sum of the reciprocals of all capacitances

Explanation: The total capacitance of capacitors connected in series is calculated using the reciprocal of the sum of the reciprocals of each capacitor's capacitance.

Correct Answer 95: B. Voltage drops due to sudden high current draw

Explanation: Flickering lights when large appliances are turned on are usually caused by voltage drops in the circuit due to the high current draw of the appliances momentarily reducing the available voltage for lighting.

Correct Answer 96: B. 120V

Explanation: The standard domestic voltage supply in the United States is 120V, which is the nominal voltage for most household appliances and lighting.

Correct Answer 97: D. Decreasing temperature

Explanation: The resistance of a conductor decreases when its temperature is decreased, which enhances conductivity by allowing electrons to move more freely through the conductor.

Correct Answer 98: B. To control a large current with a smaller current

Explanation: A relay is used in electrical circuits to control large currents with smaller currents by acting as a switch that is operated electrically, allowing for circuit control without direct contact.

Correct Answer 99: C. Ground Fault Circuit Interrupter (GFCI)

Explanation: A GFCI is used to interrupt the power when it detects an imbalance between incoming and outgoing current, which typically indicates a leakage current possibly due to a ground fault, enhancing safety.

Correct Answer 100: B. Low-voltage applications

Explanation: In low-voltage applications, the neutral wire must be insulated to ensure safety and prevent any accidental contact that could lead to electrical shocks or short circuits.

Correct Answer 101: D. All of the above

Explanation: A phase-to-phase fault in a power system can be caused by insulation breakdown, excessive load, improper grounding, or a combination of these factors, leading to dangerous short circuits.

Correct Answer 102: C. Lead-Acid

Explanation: Lead-acid batteries are commonly used in automotive applications due to their ability to provide high surge currents and reliability at a relatively low cost.

Correct Answer 103: B. To provide a path for fault current back to the earth

Explanation: The main purpose of using a ground wire in electrical installations is to provide a safe path for fault currents to travel back to the earth, preventing dangerous voltages from appearing on equipment casings.

Correct Answer 104: B. It increases the total capacitance

Explanation: Adding capacitors in parallel increases the total capacitance because the effective plate area increases, allowing more charge to be stored at the same voltage.

Electrical Calculations

Q105: What is the formula to calculate the total resistance of resistors connected in series?

A. $R_{total} = R_1 + R_2 + \ldots + R_n$
B. $R_{total} = \frac{1}{R_1} + \frac{1}{R_2} + \ldots + \frac{1}{R_n}$
C. $R_{total} = \frac{R_1 \times R_2 \times \ldots \times R_n}{R_1 + R_2 + \ldots + R_n}$
D. $R_{total} = R_1 \times R_2 \times \ldots \times R_n$

Q106: How do you calculate the total capacitance of capacitors connected in parallel?

A. $C_{total} = C_1 + C_2 + \ldots + C_n$
B. $C_{total} = \frac{1}{C_1} + \frac{1}{C_2} + \ldots + \frac{1}{C_n}$
C. $C_{total} = \frac{C_1 \times C_2 \times \ldots \times C_n}{C_1 + C_2 + \ldots + C_n}$
D. $C_{total} = C_1 \times C_2 \times \ldots \times C_n$

Q107: What is the voltage drop across a 10 ohm resistor carrying 5 amps of current?

A. 2 volts
B. 5 volts
C. 10 volts
D. 50 volts

Q108: How is electrical power calculated when you know the resistance and current in a circuit?

A. $P = V \times I$
B. $P = I^2 \times R$
C. $P = V^2 \times R$
D. $P = \frac{V^2}{R}$

Q109: What formula is used to calculate the current in a circuit when the voltage and resistance are known?

A. $I = \frac{V}{R}$
B. $I = V \times R$
C. $I = \frac{R}{V}$
D. $I = V + R$

Q110: In a three-phase system, if each phase has a voltage of 120V, what is the line-to-line voltage in a wye configuration?

A. 120 volts
B. 208 volts
C. 240 volts
D. 360 volts

Q111: How do you calculate the power factor in a circuit if you know the real power and the apparent power?

A. $\text{Power Factor} = \frac{\text{Real Power}}{\text{Apparent Power}}$
B. $\text{Power Factor} = \frac{\text{Apparent Power}}{\text{Real Power}}$
C. $\text{Power Factor} = \text{Real Power} \times \text{Apparent Power}$
D. $\text{Power Factor} = \sqrt{\text{Real Power}^2 + \text{Apparent Power}^2}$

Q112: What is the required wire size for a circuit carrying 20 amps with a total length of 50 feet, to keep voltage drop under 3% for a 120V system?

A. 14 AWG
B. 12 AWG
C. 10 AWG
D. 8 AWG

Q113: How do you calculate the total resistance in a parallel circuit?

A. $R_{total} = \frac{1}{\left(\frac{1}{R_1} + \frac{1}{R_2} + \ldots + \frac{1}{R_n}\right)}$
B. $R_{total} = R_1 + R_2 + \ldots + R_n$
C. $R_{total} = \frac{R_1 \times R_2 \times \ldots \times R_n}{R_1 + R_2 + \ldots + R_n}$
D. $R_{total} = R_1 \times R_2 \times \ldots \times R_n$

Q114: What is the current flowing through a 240V, 4800 watt appliance?

A. 10 amps
B. 20 amps

C. 30 amps
D. 40 amps

Q115: What is the effective resistance of three 6-ohm resistors connected in parallel?

A. 2 ohms
B. 3 ohms
C. 6 ohms
D. 18 ohms

Q116: Calculate the power dissipated by a 250 ohm resistor when a current of 2 amps flows through it.

A. 125 watts
B. 250 watts
C. 500 watts
D. 1000 watts

Q117: What is the impedance of a circuit that has a resistance of 8 ohms and a reactance of 6 ohms?

A. 2 ohms
B. 7 ohms
C. 10 ohms
D. 14 ohms

Q118: If the line voltage in a delta system is 480 volts, what is the phase voltage?

A. 240 volts
B. 277 volts
C. 480 volts
D. 960 volts

Q119: Calculate the capacitance required to achieve a power factor of 1.0 in an AC circuit consuming 5000 VA with a power factor of 0.8.

A. microfarads
B. 15 microfarads
C. 150 microfarads
D. 1500 microfarads

Q120: How do you determine the voltage at the end of a long cable run if you know the voltage drop percentage and the initial voltage?

A. Multiply the initial voltage by the voltage drop percentage
B. Subtract the voltage drop (in volts) from the initial voltage
C. Divide the initial voltage by the voltage drop percentage
D. Add the voltage drop (in volts) to the initial voltage

Q121: What is the total inductance of two 4-mH inductors connected in series?

A. 2 mH
B. 4 mH

C. 8 mH
D. 16 mH

Q122: How do you calculate the phase angle in an AC circuit if the resistance is 10 ohms and the reactance is 10 ohms?

A. 45 degrees
B. 50 degrees
C. 55 degrees
D. 60 degrees

Q123: What is the required size of a fuse for a circuit with a maximum current of 25 amps?

A. 20 amp fuse
B. 25 amp fuse
C. 30 amp fuse
D. 35 amp fuse

Q124: How do you calculate the equivalent series resistance of two resistors of 10 ohms and 20 ohms connected in series?

A. 5 ohms
B. 10 ohms
C. 15 ohms
D. 30 ohms

Q125: What is the maximum allowable current for a 12-gauge copper wire according to the NEC?

A. 15 amps
B. 20 amps
C. 25 amps
D. 30 amps

Q126: How do you calculate the kVA of a transformer if the primary voltage is 5000 volts and the secondary current is 50 amps?

A. 0.5 kVA
B. 5 kVA
C. 25 kVA
D. 250 kVA

Q127: If a motor has a power rating of 1 kW and operates at 80% efficiency, what is the actual power usage?

A. 800 watts
B. 1000 watts
C. 1250 watts
D. 2000 watts

Q128: Calculate the total reactance of two inductors of 3 ohms and 4 ohms connected in series.

A. 1 ohm
B. 3.5 ohms

C. 7 ohms
D. 12 ohms

Q129: What is the voltage across a 50-ohm resistor if 2 amps of current are flowing through it?

A. 25 volts
B. 50 volts
C. 100 volts
D. 200 volts

Q130: How much energy does a 60-watt bulb consume if left on for 10 hours?

A. 0.6 kWh
B. 6 kWh
C. 60 kWh
D. 600 kWh

Q131: What is the total resistance of three 3-ohm resistors connected in parallel?

A. 1 ohm
B. 3 ohms
C. 9 ohms
D. 27 ohms

Q132: Calculate the power dissipated in a circuit with a voltage of 12 volts and a current of 2 amperes.

A. 6 watts
B. 12 watts
C. 24 watts
D. 48 watts

Q133: If the resistance in a circuit is doubled while the voltage remains constant, what happens to the current?

A. It doubles
B. It halves
C. It quadruples
D. It remains the same

Q134: How do you calculate the total inductance of two inductors of 5 mH each connected in parallel?

A. 2.5 mH
B. 5 mH
C. 10 mH
D. 20 mH

Q135: What is the phase difference between current and voltage in an ideal capacitor?

A. 0 degrees
B. 45 degrees
C. 90 degrees
D. 180 degrees

Q136: What formula is used to calculate the reactance of a capacitor at a given frequency?

A. $X_C = \frac{1}{2\pi fC}$
B. $X_C = 2\pi fC$
C. $X_C = \frac{1}{\pi fC}$
D. $X_C = \pi fC$

Q137: Calculate the voltage drop across a resistor of 10 ohms carrying a current of 5 amperes.

A. 2 volts
B. 15 volts
C. 50 volts
D. 100 volts

Q138: What is the effective resistance of two 12-ohm resistors connected in parallel?

A. 6 ohms
B. 12 ohms
C. 24 ohms
D. None of the above

Q139: How much current flows through a 10-ohm resistor with a 100-volt source across it?

A. 1 amp
B. 10 amps
C. 100 amps
D. 1000 amps

Q140: Calculate the total capacitance of two capacitors of 100 microfarads and 200 microfarads connected in series.

A. 66.67 microfarads
B. 100 microfarads
C. 150 microfarads
D. 300 microfarads

Q141: What is the current draw of a 1.5 kW heater operating at 120 volts?

A. 12.5 amps
B. 15 amps
C. 20 amps
D. 25 amps

Q142: If the power factor of a circuit is 0.8, what is the real power if the apparent power is 1000 VA?

A. 800 watts
B. 1000 watts
C. 1250 watts
D. 1600 watts

Q143: Calculate the reactance of an inductor with a value of 10 mH at a frequency of 50 Hz.

A. 0.5 ohms

B. 3.14 ohms
C. 6.28 ohms
D. 10 ohms

Q144: What is the resistance of a heater that consumes 2000 watts at 240 volts?

A. 12 ohms
B. 28.8 ohms
C. 48 ohms
D. 72 ohms

Q145: How do you find the power dissipated by a component if you know the voltage across it and the current flowing through it?

A. Multiply the current by the voltage
B. Divide the voltage by the current
C. Add the voltage and current
D. Subtract the voltage from the current

Q146: What is the impedance of a circuit with a resistor of 4 ohms and an inductor of 3 ohms reactance in series?

A. 1 ohm
B. 5 ohms
C. 7 ohms
D. 12 ohms

Q147: How much energy is consumed by a 100-watt light bulb running for 8 hours?

A. 0.8 kWh
B. 8 kWh
C. 80 kWh
D. 800 kWh

Q148: Calculate the total resistance for a combination of four 4-ohm resistors in series.

A. 1 ohm
B. 4 ohms
C. 16 ohms
D. 64 ohms

Q149: What voltage is needed to produce 10 amps of current through a 50-ohm resistor?

A. 5 volts
B. 50 volts
C. 500 volts
D. 5000 volts

Q150: If a circuit with a total resistance of 25 ohms is drawing a current of 2 amperes, what is the power consumed?

A. 25 watts
B. 50 watts
C. 100 watts
D. 200 watts

Q151: Calculate the capacitance needed to store 10 coulombs of charge at 10 volts.

A. 0.1 farads
B. 1 farad
C. 10 farads
D. 100 farads

Q152: What is the total reactance of two 3-ohm capacitors and one 6-ohm inductor connected in series?

A. 12 ohms
B. 9 ohms
C. 6 ohms
D. 3 ohms

Q153: How is the current calculated in a circuit with three resistors of 10 ohms each, connected in parallel and a voltage supply of 120 volts?

A. 4 amps
B. 12 amps
C. 36 amps
D. 120 amps

Q154: What is the phase angle for a circuit with equal values of resistance and reactance?

A. 30 degrees
B. 45 degrees
C. 60 degrees
D. 90 degrees

Q155: If you double the voltage in a circuit while the resistance remains constant, what happens to the current?

A. It halves
B. It remains the same
C. It doubles
D. It quadruples

Q156: Calculate the voltage across a resistor of 100 ohms with a current of 0.5 amps flowing through it.

A. 20 volts
B. 50 volts
C. 100 volts
D. 200 volts

Answers

Answer 105: A. $R_{total} = R_1 + R_2 + \ldots + R_n$

Explanation: The total resistance of resistors connected in series is the sum of the resistances of all the resistors in the series.

Answer 106: A. $(C_{total} = C_1 + C_2 + \ldots + C_n)$

Explanation: The total capacitance of capacitors connected in parallel is the sum of the capacitances of each capacitor.

Answer 107: D. 50 volts

Explanation: The voltage drop across a resistor in a circuit is calculated using Ohm's Law, $(V = IR)$, where $(I = 5)$ amps and $(R = 10)$ ohms, thus $(V = 5 \times 10 = 50)$ volts.

Answer 108: B. $(P = I^2 \times R)$

Explanation: The power in a circuit, when resistance and current are known, can be calculated using $(P = I^2 \times R)$.

Answer 109: A. $(I = \frac{V}{R})$

Explanation: The current in a circuit can be calculated by dividing the voltage by the resistance, according to Ohm's Law.

Answer 110: B. 208 volts

Explanation: In a wye configuration, the line-to-line voltage is $(\sqrt{3})$ times the phase voltage, thus $(\sqrt{3} \times 120 \approx 208)$ volts.

Answer 111: A. $(\text{Power Factor} = \frac{\text{Real Power}}{\text{Apparent Power}})$

Explanation: Power factor is calculated as the ratio of real power to apparent power.

Answer 112: C. 10 AWG

Explanation: For a circuit carrying 20 amps and aiming to keep the voltage drop under 3% for a 120V system, a 10 AWG wire is typically suitable, depending on the distance and specific installation conditions.

Answer 113: A. $(R_{total} = \frac{1}{\left(\frac{1}{R_1} + \frac{1}{R_2} + \ldots + \frac{1}{R_n}\right)})$

Explanation: The total resistance of resistors connected in parallel is calculated by the reciprocal of the sum of the reciprocals of each resistor's resistance.

Answer 114: B. 20 amps

Explanation: The current through an appliance can be calculated by dividing the power by the voltage, $(I = \frac{P}{V} = \frac{4800}{240} = 20)$ amps.

Answer 115: A. 2 ohms

Explanation: The effective resistance of resistors connected in parallel is calculated by the formula $R_{total} = \frac{1}{\left(\frac{1}{R_1} + \frac{1}{R_2} + \ldots + \frac{1}{R_n}\right)}$. For three 6-ohm resistors, $R_{total} = \frac{1}{\left(\frac{1}{6} + \frac{1}{6} + \frac{1}{6}\right)} = 2$ ohms.

Answer 116: C. 500 watts

Explanation: Power dissipated by a resistor can be calculated using $P = I^2 \times R$, where $I = 2$ amps and $R = 250$ ohms. Thus, $P = 2^2 \times 250 = 500$ watts.

Answer 117: C. 10 ohms

Explanation: The impedance of a circuit with resistance and reactance can be calculated using $Z = \sqrt{R^2 + X^2}$, where $R = 8$ ohms and $X = 6$ ohms. Thus, $Z = \sqrt{8^2 + 6^2} = 10$ ohms.

Answer 118: A. 240 volts

Explanation: In a delta configuration, the phase voltage is equal to the line voltage, so it remains 240 volts.

Answer 119: C. 150 microfarads

Explanation: To adjust the power factor to 1.0, capacitive reactance needs to counteract the inductive reactance present, which requires calculating the change in reactive power and selecting a capacitor to provide this at the operating frequency.

Answer 120: B. Subtract the voltage drop (in volts) from the initial voltage

Explanation: To find the end voltage after a known voltage drop percentage, calculate the voltage drop in volts (percentage of the initial voltage) and subtract from the initial voltage.

Answer 121: C. 8 mH

Explanation: The total inductance of inductors in series is the sum of the inductances, thus $4 \text{ mH} + 4 \text{ mH} = 8 \text{ mH}$.

Answer 122: A. 45 degrees

Explanation: The phase angle θ in an AC circuit can be calculated when resistance and reactance are equal using $\theta = \tan^{-1}\left(\frac{X}{R}\right)$, resulting in 45 degrees.

Answer 123: C. 30 amp fuse

Explanation: The fuse rating should generally be higher than the maximum current to prevent nuisance blowing, yet close enough to protect the circuit effectively.

Answer 124: D. 30 ohms

Explanation: The equivalent series resistance of two resistors in series is the sum of their resistances, $10 \text{ ohms} + 20 \text{ ohms} = 30 \text{ ohms}$.

Answer 125: B. 20 amps

Explanation: According to the National Electrical Code, a 12-gauge copper wire is rated for up to 20 amps under typical conditions.

Answer 126: C. 25 kVA

Explanation: The kVA of a transformer is calculated by $\text{kVA} = \frac{\text{Voltage} \times \text{Current}}{1000}$, thus $\text{kVA} = \frac{5000 \times 50}{1000} = 25 \text{ kVA}$.

Here are the correct answers and explanations for the electrical calculation questions:

Answer 127: C. 1250 watts

Explanation: The actual power usage of a motor that operates at 80% efficiency and has a power rating of 1 kW (1000 watts) is calculated by dividing the power rating by the efficiency: $\frac{1000 \text{ W}}{0.8} = 1250 \text{ W}$.

Answer 128: C. 7 ohms

Explanation: The total reactance of two inductors connected in series is the sum of their reactances: $3 \text{ ohms} + 4 \text{ ohms} = 7 \text{ ohms}$.

Answer 129: C. 100 volts

Explanation: The voltage across a resistor can be calculated using Ohm's Law, $V = IR$, where $I = 2 \text{ amps}$ and $R = 50 \text{ ohms}$, thus $V = 2 \times 50 = 100 \text{ volts}$.

Answer 130: B. 6 kWh

Explanation: The energy consumption is calculated by multiplying the power rating by the time: $60 \text{ watts} \times 10 \text{ hours} = 600 \text{ watt-hours} = 0.6 \text{ kWh}$.

Answer 131: A. 1 ohm

Explanation: The total resistance of resistors connected in parallel is calculated by the reciprocal of the sum of the reciprocals of each resistor's resistance. For three 3-ohm resistors: $R_{total} = \frac{1}{(\frac{1}{3}+\frac{1}{3}+\frac{1}{3})} = 1 \text{ ohm}$.

Answer 132: C. 24 watts

Explanation: Power in a circuit can be calculated using the formula $P = VI$, where $V = 12 \text{ volts}$ and $I = 2 \text{ amperes}$, thus $P = 12 \times 2 = 24 \text{ watts}$.

Answer 133: B. It halves

Explanation: If the resistance is doubled and the voltage remains constant, the current will halve according to Ohm's Law ($I = \frac{V}{R}$).

Answer 134: A. 2.5 mH

Explanation: The total inductance of inductors connected in parallel is calculated using the formula for parallel resistors but applied to inductance: $L_{total} = \frac{1}{(\frac{1}{5}+\frac{1}{5})} = 2.5 \text{ mH}$.

Answer 135: C. 90 degrees

Explanation: In an ideal capacitor, the phase difference between current and voltage is 90 degrees, with the current leading the voltage.

Answer 136: A. $X_C = \frac{1}{2\pi fC}$

Explanation: The reactance of a capacitor at a given frequency is calculated using the formula $X_C = \frac{1}{2\pi fC}$.

Answer 137: C. 50 volts

Explanation: The voltage drop across a resistor is calculated using Ohm's Law, $V = IR$, where $I = 5$ amps and $R = 10$ ohms, thus $V = 5 \times 10 = 50$ volts.

Answer 138: A. 6 ohms

Explanation: The effective resistance of two 12-ohm resistors connected in parallel is calculated by the reciprocal of the sum of the reciprocals formula: $R_{total} = \frac{1}{(\frac{1}{12}+\frac{1}{12})} = 6 \text{ ohms}$.

Answer 139: B. 10 amps

Explanation: The current through a resistor can be calculated by $I = \frac{V}{R}$, where $V = 100 \text{ volts}$ and $R = 10 \text{ ohms}$, thus $I = \frac{100}{10} = 10 \text{ amps}$.

Answer 140: A. 66.67 microfarads

Explanation: The total capacitance of capacitors connected in series is calculated using the formula $C_{total} = \frac{1}{(\frac{1}{100}+\frac{1}{200})} = 66.67 \text{ microfarads}$.

Answer 141: A. 12.5 amps

Explanation: The current draw of a device can be calculated using $I = \frac{P}{V}$, where $P = 1.5 \text{ kW} = 1500 \text{ watts}$ and $V = 120 \text{ volts}$, thus $I = \frac{1500}{120} = 12.5 \text{ amps}$.

Answer 142: A. 800 watts

Explanation: Real power is calculated by multiplying the power factor by the apparent power: $\text{Real Power} = 0.8 \times 1000 \text{ VA} = 800 \text{ watts}$.

Answer 143: B. 3.14 ohms

Explanation: The reactance of an inductor is calculated using $X_L = 2\pi fL$, where $f = 50 \text{ Hz}$ and $L = 10 \text{ mH} = 0.01 \text{ H}$, thus $X_L = 2\pi \times 50 \times 0.01 = 3.14 \text{ ohms}$.

Answer 144: B. 28.8 ohms

Explanation: The resistance of a heater can be calculated using $R = \frac{V^2}{P}$, where $P = 2000 \text{ watts}$ and $V = 240 \text{ volts}$, thus $R = \frac{240^2}{2000} = 28.8 \text{ ohms}$.

Answer 145: A. Multiply the current by the voltage

Explanation: Power dissipated by a component is calculated using $P = VI$, where V is the voltage across the component and I is the current flowing through it.

Answer 146: C. 7 ohms

Explanation: The impedance of a series circuit with resistance and reactance is calculated using $Z = \sqrt{R^2 + X^2}$, where $R = 4$ ohms and $X = 3$ ohms, thus $Z = \sqrt{4^2 + 3^2} = 7$ ohms.

Answer 147: B. 8 kWh

Explanation: Energy consumption is calculated by multiplying the power by the time: $100 \text{ watts} \times 8 \text{ hours} = 800 \text{ watt-hours} = 0.8 \text{ kWh}$.

Answer 148: C. 16 ohms

Explanation: The total resistance for resistors in series is the sum of their resistances: $4 \text{ ohms} + 4 \text{ ohms} + 4 \text{ ohms} + 4 \text{ ohms} = 16 \text{ ohms}$.

Answer 149: C. 500 volts

Explanation: The voltage needed to produce a certain current through a resistor is calculated using Ohm's Law, $V = IR$, where $I = 10$ amps and $R = 50 \text{ ohms}$, thus $V = 10 \times 50 = 500 \text{ volts}$.

Answer 150: C. 100 watts

Explanation: The power consumed in a circuit can be calculated using $P = I^2R$, where $I = 2$ amps and $R = 25$ ohms, thus $P = 2^2 \times 25 = 100 \text{ watts}$.

Answer 151: B. 1 farad

Explanation: Capacitance needed to store charge at a certain voltage is calculated using $C = \frac{Q}{V}$, where $Q = 10$ coulombs and $V = 10$ volts, thus $C = \frac{10}{10} = 1 \text{ farad}$.

Answer 152: C. 6 ohms

Explanation: The total reactance of components in series is the sum of their reactances, regardless of whether they are capacitive or inductive: $3 \text{ ohms} + 3 \text{ ohms} + 6 \text{ ohms} = 12 \text{ ohms}$.

Answer 153: C. 36 amps

Explanation: For resistors in parallel, $R_{total} = \frac{1}{(\frac{1}{10}+\frac{1}{10}+\frac{1}{10})} = 3.33 \text{ ohms}$. The current through the circuit is $I = \frac{V}{R} = \frac{120}{3.33} \approx 36 \text{ amps}$.

Answer 154: B. 45 degrees

Explanation: For circuits with equal values of resistance and reactance, the phase angle θ is 45 degrees.

Answer 155: C. It doubles

Explanation: If voltage is doubled while resistance remains constant, current also doubles according to Ohm's Law ($I = \frac{V}{R}$).

Answer 156: B. 50 volts

Explanation: Voltage across a resistor is calculated using Ohm's Law, $V = IR$, where $I = 0.5$ amps and $R = 100$ ohms, thus $V = 0.5 \times 100 = 50$ volts.

Expertise in Wiring Methods and Materials

Q157: What is the total impedance in a circuit where a 50-ohm resistor and a 50-ohm capacitive reactance are connected in parallel?

A. 25 ohms
B. 50 ohms
C. 100 ohms
D. None of the above

Q158: Calculate the power dissipated in a resistor of 25 ohms when a current of 4 amps flows through it.

A. 100 watts
B. 200 watts
C. 400 watts
D. 600 watts

Q159: What is the voltage across a 200-ohm resistor if it draws a current of 0.5 amps?

A. 100 volts
B. 200 volts
C. 400 volts
D. 500 volts

Q160: How do you calculate the total capacitance of capacitors connected in series when given two capacitors, each with a capacitance of 10 microfarads?

A. 5 microfarads
B. 10 microfarads
C. 20 microfarads
D. 50 microfarads

Q161: What is the resistance of a wire that allows a current of 10 amps to flow when connected across a 120-volt source?

A. 12 ohms
B. 15 ohms
C. 20 ohms
D. 24 ohms

Q162: Calculate the total inductance when two inductors, each with an inductance of 5 mH, are connected in parallel.

A. 2.5 mH
B. 5 mH
C. 10 mH
D. 20 mH

Q163: If a circuit's power factor is 0.75 and the real power is 300 watts, what is the apparent power?

A. 225 VA
B. 400 VA
C. 450 VA
D. 600 VA

Q164: What is the phase difference in an AC circuit with only capacitive load?

A. 0 degrees
B. 45 degrees
C. 90 degrees
D. 180 degrees

Q165: How much current flows through a circuit with a voltage of 120 volts and a resistance of 24 ohms?

A. 0.2 amps
B. 2.5 amps
C. 5 amps
D. 10 amps

Q166: Calculate the energy used by a 150-watt light bulb that runs for 12 hours.

A. 1.8 kWh
B. 1800 kWh
C. 18000 kWh
D. 180000 kWh

Q167: What is the total resistance in a parallel circuit with three resistors of 30 ohms, 60 ohms, and 90 ohms?

A. 10 ohms
B. 15 ohms
C. 20 ohms
D. 30 ohms

Q168: Calculate the voltage drop across a 10 ohm resistor when a current of 3 amps flows through it.

A. 3 volts
B. 10 volts
C. 30 volts
D. 300 volts

Q169: What is the impedance of a circuit with a 10 ohm resistor in series with a 10 ohm inductor?

A. 10 ohms
B. 14.14 ohms
C. 20 ohms
D. 30 ohms

Q170: How do you calculate the reactance of an inductor with an inductance of 2 mH at a frequency of 1000 Hz?

A. 0.2 ohms
B. 2 ohms
C. 6.28 ohms
D. 12.57 ohms

Q171: What is the total power consumption of three devices that consume 100 watts, 200 watts, and 300 watts respectively when connected in parallel to a voltage supply?

A. 100 watts
B. 200 watts
C. 300 watts
D. 600 watts

Q172: Calculate the required capacitance to correct the power factor from 0.7 to

1.0 for a circuit consuming 500 watts at 120 volts.

A. 183 microfarads
B. 366 microfarads
C. 549 microfarads
D. 732 microfarads

Q173: What is the current flow through a diode when a forward bias voltage of 0.7 volts is applied, and the resistance is 700 ohms?

A. 0.001 amps
B. 0.01 amps
C. 1 amp
D. 10 amps

Q174: Calculate the equivalent resistance of a network with two parallel branches, one with a resistance of 50 ohms and the other with 100 ohms.

A. 25 ohms
B. 33.33 ohms
C. 50 ohms
D. 150 ohms

Q175: What voltage is required to pass a current of 2 amps through a resistor of 100 ohms?

A. 50 volts
B. 100 volts
C. 200 volts
D. 400 volts

Q176: How much power is dissipated by a resistor of 20 ohms when a current of 2 amps flows through it?

A. 20 watts
B. 40 watts
C. 80 watts
D. 160 watts

Q177: Calculate the voltage across a resistor of 5 ohms when it is in a series circuit with a total current of 10 amps.

A. 2 volts
B. 5 volts
C. 50 volts
D. 100 volts

Q178: What is the total capacitance of three capacitors of 100 microfarads each connected in series?

A. 33.33 microfarads
B. 100 microfarads
C. 300 microfarads
D. 900 microfarads

Q179: If the line voltage in a delta system is 480 volts, what is the phase current if the phase resistance is 30 ohms?

A. 8 amps
B. 16 amps
C. 32 amps
D. 64 amps

Q180: Calculate the inductance required to achieve a reactance of 50 ohms at a frequency of 60 Hz.

A. 0.132 mH
B. 1.32 mH
C. 13.2 mH
D. 132 mH

Q181: What is the voltage drop across a 10-ohm resistor if the power dissipated is 100 watts?

A. 10 volts
B. 31.6 volts
C. 100 volts
D. 316 volts

Q182: How much energy does a 75-watt bulb consume if it runs for 24 hours?

A. 1.8 kWh
B. 18 kWh
C. 180 kWh
D. 1800 kWh

Q183: Calculate the current through a circuit with a voltage supply of 240 volts and a resistance of 60 ohms.

A. 2 amps
B. 4 amps
C. 6 amps
D. 8 amps

Q184: What is the total resistance of two 20-ohm resistors connected in series?

A. 10 ohms
B. 20 ohms
C. 40 ohms
D. 80 ohms

Q185: Calculate the power dissipated in a resistor of 100 ohms when a current of 3 amps flows through it.

A. 300 watts
B. 600 watts
C. 900 watts
D. 1200 watts

Q186: What is the voltage across a capacitor in an RC circuit if the charge on the capacitor is 2 coulombs and the capacitance is 4 farads?

A. 0.5 volts
B. 2 volts
C. 8 volts
D. 16 volts

Q187: How do you calculate the power factor if the real power is 400 watts and the apparent power is 500 VA?

A. 0.8
B. 0.5
C. 1.25
D. 2

Q188: What is the phase angle between voltage and current in a circuit where the resistance is equal to the reactance?

A. 30 degrees
B. 45 degrees
C. 60 degrees
D. 90 degrees

Q189: Calculate the resistance required to produce a voltage drop of 50 volts when a current of 5 amps flows through it.

A. 5 ohms
B. 10 ohms
C. 25 ohms
D. 50 ohms

Q190: What is the total inductance of two 5 mH inductors connected in series?

A. 2.5 mH
B. 5 mH
C. 10 mH
D. 20 mH

Q191: How much current will flow through a circuit with a voltage supply of 110 volts and a resistance of 22 ohms?

A. 2 amps
B. 5 amps
C. 10 amps
D. 20 amps

Q192: Calculate the total capacitance when three 30 microfarad capacitors are connected in series.

A. 10 microfarads
B. 30 microfarads
C. 90 microfarads
D. 270 microfarads

Q193: What is the voltage drop across a 30-ohm resistor if a current of 2 amps flows through it?

A. 15 volts
B. 30 volts
C. 60 volts
D. 120 volts

Q194: How do you calculate the reactance of a 5 mH inductor at a frequency of 100 Hz?

A. 0.314 ohms
B. 3.14 ohms
C. 31.4 ohms
D. 314 ohms

Q195: Calculate the effective resistance of two 8-ohm speakers connected in parallel.

A. 4 ohms
B. 8 ohms
C. 16 ohms
D. 32 ohhs

Q196: What is the power consumption of a device that operates at 240 volts with a current of 2 amps?

A. 120 watts
B. 240 watts
C. 480 watts
D. 960 watts

Q197: How much energy is consumed by a device rated at 500 watts operating for 3 hours?

A. 1.5 kWh
B. 500 kWh
C. 1500 kWh
D. 15000 kWh

Q198: What is the total impedance in a circuit where a 100-ohm resistor, a 100-ohm inductive reactance, and a 100-ohm capacitive reactance are connected in series?

A. 0 ohms
B. 100 ohms
C. 300 ohms
D. None of the above

Q199: Calculate the phase angle in an AC circuit where the inductive reactance is 50 ohms and the resistance is 50 ohms.

A. 30 degrees
B. 45 degrees
C. 60 degrees
D. 90 degrees

Q200: What voltage is needed to drive a current of 10 amps through a resistance of 50 ohms?

A. 5 volts
B. 50 volts
C. 500 volts
D. 5000 volts

Q201: Calculate the total capacitance for four 100 microfarad capacitors connected in parallel.

A. 25 microfarads
B. 100 microfarads
C. 400 microfarads
D. 1600 microfarads

Q202: What is the current through a 200-watt bulb operating at 100 volts?

A. 0.5 amps
B. 2 amps
C. 5 amps
D. 20 amps

Q203: How do you calculate the total resistance of a resistor network if two resistors of 10 ohms each are connected in parallel?

A. 5 ohms
B. 10 ohms
C. 20 ohms
D. None of the above

108

Q204: Calculate the reactance of a capacitor with a capacitance of 10 microfarads at a frequency of 1000 Hz.

A. 1.59 ohms
B. 15.9 ohms
C. 159 ohms
D. 1590 ohms

Q205: What is the phase difference between voltage and current in an ideal inductor?

A. 0 degrees
B. 45 degrees
C. 90 degrees
D. 180 degrees

Q206: Calculate the total power dissipated in a circuit with three resistors of 5 ohms, 10 ohms, and 20 ohms connected in series, with a current of 2 amps flowing through them.

A. 10 watts
B. 70 watts
C. 140 watts
D. 280 watts

Answers

Answer 157: D. None of the above

Explanation: The total impedance in a parallel circuit with a resistor and a capacitive reactance both valued at 50 ohms can be calculated using the formula for parallel resistances: $Z_{total} = \frac{1}{\sqrt{(\frac{1}{R})^2 + (\frac{1}{X_C})^2}}$. This results in an impedance that is not one of the provided options.

Answer 158: C. 400 watts

Explanation: Power dissipated in a resistor is calculated using $P = I^2R$, where $I = 4$ amps and $R = 25$ ohms. Therefore, $P = 4^2 \times 25 = 400$ watts.

Answer 159: A. 100 volts

Explanation: The voltage across a resistor is calculated using Ohm's Law, $V = IR$, where $I = 0.5$ amps and $R = 200$ ohms. Therefore, $V = 0.5 \times 200 = 100$ volts.

Answer 160: A. 5 microfarads

Explanation: The total capacitance of capacitors in series is calculated using the formula $C_{total} = \frac{1}{(\frac{1}{C_1} + \frac{1}{C_2})}$, where each capacitor is 10 microfarads. Therefore, $C_{total} = \frac{1}{(\frac{1}{10} + \frac{1}{10})} = 5$ microfarads.

Answer 161: A. 12 ohms

Explanation: The resistance of a wire allowing a current flow can be calculated using Ohm's Law, $R = \frac{V}{I}$, where $V = 120$ volts and $I = 10$ amps. Therefore, $R = \frac{120}{10} = 12$ ohms.

Answer 162: A. 2.5 mH

Explanation: The total inductance of inductors connected in parallel is calculated using $L_{total} = \frac{1}{(\frac{1}{L_1} + \frac{1}{L_2})}$, where each inductor is 5 mH. Therefore, $L_{total} = \frac{1}{(\frac{1}{5} + \frac{1}{5})} = 2.5$ mH.

Answer 163: B. 400 VA

Explanation: The apparent power in a circuit can be calculated as $\text{Apparent Power} = \frac{\text{Real Power}}{\text{Power Factor}}$, where the real power is 300 watts and the power factor is 0.75. Therefore, $\text{Apparent Power} = \frac{300}{0.75} = 400$ VA.

Answer 164: C. 90 degrees

Explanation: In an AC circuit with only capacitive load, the current leads the voltage by 90 degrees.

Answer 165: C. 5 amps

Explanation: The current through a circuit is calculated using Ohm's Law, $I = \frac{V}{R}$, where $V = 120$ volts and $R = 24$ ohms. Therefore, $I = \frac{120}{24} = 5$ amps.

Answer 166: A. 1.8 kWh

Explanation: The energy used is calculated by multiplying the power by the time in hours, $150 \text{ watts} \times 12 \text{ hours} = 1800 \text{ watt-hours} = 1.8 \text{ kWh}$.

Answer 167: C. 20 ohms

Explanation: The total resistance for resistors in parallel is calculated using the reciprocal sum formula: $R_{total} = \frac{1}{(\frac{1}{30} + \frac{1}{60} + \frac{1}{90})} = 20$ ohms.

Answer 168: C. 30 volts

Explanation: The voltage drop across a resistor is calculated using Ohm's Law, $V = IR$, where $I = 3$ amps and $R = 10$ ohms. Therefore, $V = 3 \times 10 = 30$ volts.

Answer 169: B. 14.14 ohms

Explanation: The total impedance of a circuit with a resistor and inductor in series is calculated using $Z = \sqrt{R^2 + X_L^2}$, where $R = X_L = 10$

ohms. Therefore, $Z = \sqrt{10^2 + 10^2} = 14.14$ ohms.

Answer 170: C. 6.28 ohms

Explanation: The reactance of an inductor is calculated using $X_L = 2\pi fL$, where $f = 1000$ Hz and $L = 2$ mH (0.002 H). Therefore, $X_L = 2\pi \times 1000 \times 0.002 = 6.28$ ohms.

Answer 171: D. 600 watts

Explanation: The total power consumption when devices are connected in parallel is the sum of their individual powers, $100 + 200 + 300 = 600$ watts.

Answer 172: C. 549 microfarads

Explanation: The required capacitance to correct the power factor involves calculating the reactive power needed to adjust from a power factor of 0.7 to 1.0, then using the formula to find the capacitance at the given voltage and frequency.

Answer 173: B. 0.01 amps

Explanation: The current through a diode with forward bias is calculated using $I = \frac{V}{R}$, where $V = 0.7$ volts and $R = 700$ ohms. Therefore, $I = \frac{0.7}{700} = 0.001$ amps, rounded to 0.01 amps for practical purposes.

Answer 174: B. 33.33 ohms

Explanation: The equivalent resistance of parallel resistors is calculated using the reciprocal sum formula: $R_{total} = \frac{1}{(\frac{1}{50} + \frac{1}{100})} = 33.33$ ohms.

Answer 175: C. 200 volts

Explanation: The voltage required to pass a current through a resistor is calculated using Ohm's Law, $V = IR$, where $I = 2$ amps and $R = 100$ ohms. Therefore, $V = 2 \times 100 = 200$ volts.

Answer 176: C. 80 watts

Explanation: Power dissipated by a resistor is calculated using $P = I^2R$, where $I = 2$ amps and $R = 20$ ohms. Therefore, $P = 2^2 \times 20 = 80$ watts.

Answer 177: C. 50 volts

Explanation: The voltage across a resistor is calculated using Ohm's Law, $V = IR$, where $I = 10$ amps and $R = 5$ ohms. Therefore, $V = 10 \times 5 = 50$ volts.

Answer 178: A. 33.33 microfarads

Explanation: The total capacitance of capacitors in series is calculated using the reciprocal sum formula: $C_{total} = \frac{1}{(\frac{1}{100} + \frac{1}{100} + \frac{1}{100})} = 33.33$ microfarads.

Answer 179: B. 16 amps

Explanation: The phase current in a delta system is calculated using Ohm's Law, $I = \frac{V}{R}$, where $V = 480$ volts and $R = 30$ ohms. Therefore, $I = \frac{480}{30} = 16$ amps.

Answer 180: C. 13.2 mH

Explanation: The inductance needed to achieve a given reactance at a specific frequency is calculated using $L = \frac{X_L}{2\pi f}$, where $X_L = 50$ ohms and $f = 60$ Hz. Therefore, $L = \frac{50}{2\pi \times 60} = 0.132$ H or 132 mH.

Answer 181: B. 31.6 volts

Explanation: The voltage across a resistor when power is known can be calculated using $V = \sqrt{PR}$, where $P = 100$ watts and $R = 10$ ohms. Therefore, $V = \sqrt{100 \times 10} = 31.6$ volts.

Answer 182: A. 1.8 kWh

Explanation: The energy consumed by a 75-watt bulb running for 24 hours is calculated by multiplying the power rating by the duration: $75 \text{ watts} \times 24 \text{ hours} = 1800 \text{ watt-hours} = 1.8 \text{ kWh}$.

Answer 183: B. 4 amps

Explanation: The current through a circuit is calculated using Ohm's Law: $I = \frac{V}{R}$, where $V = 240$ volts and $R = 60$ ohms. Therefore, $I = \frac{240}{60} = 4 \text{ amps}$.

Answer 184: C. 40 ohms

Explanation: The total resistance of two resistors connected in series is the sum of their resistances: $20 \text{ ohms} + 20 \text{ ohms} = 40 \text{ ohms}$.

Answer 185: C. 900 watts

Explanation: The power dissipated in a resistor is calculated using $P = I^2R$, where $I = 3 \text{ amps}$ and $R = 100 \text{ ohms}$. Thus, $P = 3^2 \times 100 = 900 \text{ watts}$.

Answer 186: B. 2 volts

Explanation: The voltage across a capacitor is calculated using $V = \frac{Q}{C}$, where $Q = 2 \text{ coulombs}$ and $C = 4 \text{ farads}$

$\)$. Therefore, $V = \frac{2}{4} = 0.5 \text{ volts}$.

Answer 187: A. 0.8

Explanation: Power factor is calculated as the ratio of real power to apparent power: $\text{Power Factor} = \frac{\text{Real Power}}{\text{Apparent Power}} = \frac{400 \text{ watts}}{500 \text{ VA}} = 0.8$.

Answer 188: B. 45 degrees

Explanation: When the resistance equals the reactance, the phase angle in an AC circuit is 45°.

Answer 189: B. 10 ohms

Explanation: The resistance required for a given voltage drop is calculated using Ohm's Law: $R = \frac{V}{I}$, where $V = 50 \text{ volts}$ and $I = 5 \text{ amps}$. Thus, $R = \frac{50}{5} = 10 \text{ ohms}$.

Answer 190: C. 10 mH

Explanation: The total inductance of inductors connected in series is the sum of their individual inductances: $5 \text{ mH} + 5 \text{ mH} = 10 \text{ mH}$.

Answer 191: B. 5 amps

Explanation: Current through a circuit is calculated using Ohm's Law: $I = \frac{V}{R}$, where $V = 110 \text{ volts}$ and $R = 22 \text{ ohms}$. Thus, $I = \frac{110}{22} = 5 \text{ amps}$.

Answer 192: A. 10 microfarads

Explanation: The total capacitance of capacitors connected in series is calculated using the formula: $C_{total} = \frac{1}{\left(\frac{1}{C_1} + \frac{1}{C_2} + \frac{1}{C_3}\right)}$, where each C is 30 microfarads. Thus, $C_{total} = \frac{1}{\left(\frac{1}{30} + \frac{1}{30} + \frac{1}{30}\right)} = 10 \text{ microfarads}$.

Answer 193: C. 60 volts

Explanation: The voltage drop across a resistor is calculated using Ohm's Law: $V = IR$, where $I = 2 \text{ amps}$ and $R = 30 \text{ ohms}$. Thus, $V = 2 \times 30 = 60 \text{ volts}$.

Answer 194: B. 3.14 ohms

Explanation: The reactance of an inductor at a given frequency is calculated using $X_L = 2\pi fL$, where $f = 100 \text{ Hz}$ and $L = 5 \text{ mH}$. Thus, $X_L = 2\pi \times 100 \times 0.005 = 3.14 \text{ ohms}$.

Answer 195: A. 4 ohms

Explanation: The effective resistance of two resistors connected in parallel is calculated using the reciprocal sum formula: $R_{total} = \frac{1}{\left(\frac{1}{R_1} + \frac{1}{R_2}\right)}$, where each resistor is 8 ohms. Thus, $R_{total} = \frac{1}{\left(\frac{1}{8} + \frac{1}{8}\right)} = 4 \text{ ohms}$.

Answer 196: C. 480 watts

Explanation: Power consumption is calculated using $P = VI$, where $V = 240 \text{ volts}$ and $I = 2 \text{ amps}$. Thus, $P = 240 \times 2 = 480 \text{ watts}$.

Answer 197: A. 1.5 kWh

Explanation: Energy consumption is calculated by multiplying the power rating by the time: $500 \text{ watts} \times 3 \text{ hours} = 1500 \text{ watt-hours} = 1.5 \text{ kWh}$.

Answer 198: D. None of the above

Explanation: In a circuit where a resistor, an inductive reactance, and a capacitive reactance are all equal and connected in series, the total impedance is zero due to the cancellation of the inductive and capacitive reactances.

Answer 199: B. 45 degrees

Explanation: The phase angle in an AC circuit where the inductive reactance equals the resistance is 45°.

Answer 200: C. 500 volts

Explanation: The voltage required to drive a current through a resistance is calculated using Ohm's Law: $V = IR$, where $I = 10 \text{ amps}$ and $R = 50 \text{ ohms}$. Thus, $V = 10 \times 50 = 500 \text{ volts}$.

Answer 201: C. 400 microfarads

Explanation: The total capacitance for capacitors connected in parallel is the sum of their capacitances: $100 \text{ microfarads} + 100 \text{ microfarads} + 100 \text{ microfarads} + 100 \text{ microfarads} = 400 \text{ microfarads}$.

Answer 202: B. 2 amps

Explanation: Current through a device is calculated using Ohm's Law: $I = \frac{P}{V}$, where $P = 200 \text{ watts}$ and $V = 100 \text{ volts}$. Thus, $I = \frac{200}{100} = 2 \text{ amps}$.

Answer 203: A. 5 ohms

Explanation: The total resistance of two resistors connected in parallel is calculated using the reciprocal sum formula: $R_{total} = \frac{1}{\left(\frac{1}{10} + \frac{1}{10}\right)} = 5 \text{ ohms}$.

Answer 204: C. 159 ohms

Explanation: The reactance of a capacitor at a given frequency is calculated using $X_C = \frac{1}{2\pi fC}$, where $f = 1000 \text{ Hz}$ and $C = 10 \text{ microfarads}$. Thus, $X_C = \frac{1}{2\pi \times 1000 \times 10 \times 10^{-6}} = 15.9 \text{ ohms}$.

Answer 205: C. 90 degrees

Explanation: In an ideal inductor, the phase difference between voltage and current is 90° with the current lagging behind the voltage.

Answer 206: C. 140 watts

Explanation: The total power dissipated in a series circuit is calculated by summing the power dissipated by each resistor, where each power is calculated as $P = I^2R$. Therefore, $P_{total} = 2^2 \times 5 + 2^2 \times 10 + 2^2 \times 20 = 140 \text{ watts}$.

Electrical Equipment and Devices

Q207: What is the primary function of a circuit breaker in an electrical system?

A. To step up voltage
B. To reduce voltage
C. To interrupt the electrical circuit during an overload or short circuit
D. To convert AC to DC

Q208: Which device is used to measure the flow of electrical current in a circuit?

A. Transformer
B. Voltmeter
C. Ammeter
D. Capacitor

Q209: What is the typical purpose of a relay in an electrical circuit?

A. To store energy
B. To provide high current from a low current signal
C. To reduce signal noise
D. To measure voltage

Q210: What type of switch is commonly used for changing the direction of motor rotation?

A. Toggle switch
B. Selector switch
C. Pressure switch
D. Limit switch

Q211: Which electrical device converts mechanical energy into electrical energy?

A. Motor
B. Generator
C. Capacitor
D. Resistor

Q212: What is the main function of a fuse in an electrical circuit?

A. To control the power flow
B. To protect the circuit by breaking the connection when current exceeds a certain limit
C. To increase current flow
D. To decrease resistance

Q213: What type of motor is typically used where precise speed control is necessary?

A. Synchronous motor
B. Induction motor
C. Stepper motor
D. Brushless DC motor

Q214: Which component is used to store electric charge in an electric field?

A. Resistor
B. Inductor
C. Capacitor
D. Transformer

Q215: In residential wiring, what is the primary purpose of the ground wire?

A. To carry live electrical current
B. To protect against electrical shock by providing a path for fault current
C. To increase circuit efficiency
D. To reduce power consumption

Q216: What is the principle operation of a transformer?

A. Electromagnetic induction
B. Piezoelectric effect
C. Thermoelectric effect
D. Photoelectric effect

Q217: What device is commonly used to start and stop an electric motor?

A. Relay
B. Circuit breaker
C. Starter
D. Fuse

Q218: Which type of switch is used to detect the absence or presence of an object?

A. Pressure switch
B. Limit switch
C. Proximity switch
D. Toggle switch

Q219: How does a thermocouple work?

A. By converting mechanical stress into voltage
B. By emitting light when electric current passes through it
C. By generating a voltage when two dissimilar metals are heated at one end
D. By changing resistance with temperature changes

Q220: What is the main use of a variable frequency drive (VFD)?

A. To control the power factor of industrial equipment
B. To adjust the speed of AC motors by varying the frequency of the electrical supply
C. To convert AC to DC
D. To step up or step down voltage levels

Q221: Which device protects electrical circuits from overvoltage conditions?

A. Surge protector
B. Circuit breaker
C. Voltage regulator
D. Transformer

Q222: What is the role of a rectifier in a power supply?

A. To maintain a constant output voltage
B. To convert alternating current (AC) to direct current (DC)
C. To decrease the voltage before it reaches sensitive components
D. To store excess electricity

Q223: Which electrical component is essential for starting fluorescent lamps?

A. Capacitor
B. Ballast
C. Resistor
D. Diode

Q224: What function does a diode serve in an electrical circuit?

A. To allow current to flow in both directions
B. To only allow current to flow in one direction
C. To block all electrical current
D. To increase current flow

Q225: What is a common application of a photocell in modern devices?

A. To measure current
B. To control motors

117

C. To detect light levels and adjust lighting automatically
D. To convert mechanical movement into electrical energy

Q226: Which device is used to disconnect an electrical circuit in the event of an overload manually?

A. Fuse
B. Circuit breaker
C. Switch
D. Relay

Q227: What is the main purpose of an inductor in a circuit?

A. To oppose changes in voltage
B. To oppose changes in current
C. To increase current flow
D. To store charge

Q228: What type of motor is most commonly used in HVAC systems for fan control?

A. Synchronous motor
B. Induction motor
C. Stepper motor
D. Servo motor

Q229: How does a piezoelectric sensor work?

A. By generating a voltage in response to mechanical stress
B. By changing resistance with temperature
C. By emitting light when activated
D. By converting light into electricity

Q230: What is the role of a solenoid in automotive applications?

A. To control fuel flow
B. To actuate brake systems
C. To convert electrical energy into linear motion
D. To filter electrical noise

Q231: Which type of cable is typically used for transmitting both power and data to networking hardware?

A. Coaxial cable
B. Twisted pair cable
C. Fiber optic cable
D. Ribbon cable

Q232: What is the function of a heat sink in electronic devices?

A. To store excess heat
B. To generate heat to warm device components
C. To dissipate heat to prevent overheating
D. To convert heat into electrical energy

Q233: What type of electrical device is used to ensure electrical safety by detecting ground faults?

A. Fuse
B. Ground Fault Circuit Interrupter (GFCI)
C. Resistor
D. Capacitor

Q234: In what type of electrical system is a capacitor bank typically used?

A. To store digital data
B. To increase voltage temporarily
C. To improve power factor in AC power systems
D. To convert AC to DC

Q235: What is the purpose of a busbar in electrical distribution?

A. To regulate voltage
B. To distribute power across multiple circuits
C. To convert frequencies
D. To store electrical energy

Q236: Which device is most commonly used to control the flow of large electric currents by means of small control signals?

A. Contactor
B. Transformer
C. Fuse
D. Capacitor

Q237: What is the function of a core in a transformer?

A. To insulate the windings
B. To reduce electrical noise
C. To provide a path for the magnetic flux
D. To cool the transformer oil

Q238: What does a multimeter primarily measure?

A. Weight
B. Temperature
C. Electrical properties like voltage, current, and resistance
D. Pressure

Q239: Which component is typically used in circuits to delay the flow of electricity for a set period?

A. Relay
B. Timer
C. Resistor
D. Transistor

Q240: What role does a stabilizer play in electrical systems?

A. To stabilize the physical structure of the system
B. To maintain constant voltage levels despite fluctuations in supply
C. To increase current flow
D. To decrease resistance

Q241: Which device is used to step down high voltages in power lines to usable levels in homes and businesses?

A. Generator
B. Transformer
C. Rectifier
D. Inverter

Q242: What is the main use of a Hall effect sensor in electronic devices?

A. To measure temperature
B. To detect changes in light
C. To measure magnetic fields and convert them into a digital value
D. To measure pressure changes

Q243: What kind of switch is commonly used in staircases for controlling lights from two different places?

A. Single-pole switch
B. Double-pole switch
C. Three-way switch
D. Dimmer switch

Q244: How does an ultrasonic sensor determine the distance to an object?

A. By measuring the intensity of light reflected from the object
B. By measuring the time it takes for a sound wave to return to the sensor
C. By detecting changes in magnetic fields
D. By measuring electrical resistance changes due to proximity

Q245: What is the primary function of an electrical isolator?

A. To isolate part of a circuit while it is being worked on
B. To convert energy forms
C. To reduce voltage
D. To amplify signals

Q246: Which type of battery is typically used in uninterruptible power supply (UPS) systems?

A. Alkaline battery
B. Lead-acid battery
C. Lithium-ion battery
D. Nickel-cadmium battery

Q247: What is a common application for a rheostat?

A. To increase voltage
B. To control the brightness of a light
C. To detect current flow
D. To insulate a circuit

Q248: What device prevents overcharging in a battery-powered system?

A. Charge controller
B. Relay
C. Capacitor
D. Transformer

Q249: In what device would you commonly find a bimetallic strip?

A. Transformer
B. Electric motor
C. Thermostat
D. Generator

Q250: What function does an encoder perform in industrial automation?

A. Measures and converts pressure into an electrical signal
B. Converts mechanical motion into an electrical signal to determine position
C. Regulates the flow of a liquid
D. Changes electrical energy into mechanical energy

Q251: What is the primary use of a varistor in electronic circuits?

A. To maintain constant voltage
B. To protect against voltage spikes
C. To increase resistance
D. To measure voltage

Q252: What kind of device is used to control lighting and machinery automatically based on ambient light levels?

A. Photocell
B. Capacitor
C. Resistor
D. Inductor

Q253: Which component is essential in solar power systems to convert DC to AC?

A. Inverter
B. Rectifier
C. Transformer
D. Conductor

Q254: How does a synchronous motor differ from an asynchronous motor?

A. It operates at a constant speed up to its full load
B. It varies speed according to the applied load
C. It requires an external inverter
D. It only operates on DC

Q255: What is a practical application for a piezoelectric transducer?

A. As a battery charger
B. As a signal amplifier
C. In electronic drum pads to convert drum strikes into electrical signals
D. As a voltage regulator

Q256: Which device is primarily used to measure and display electrical power consumption?

A. Wattmeter
B. Ammeter
C. Voltmeter
D. Multimeter

Q257: What type of device is a PLC (Programmable Logic Controller)?

A. A device used for high voltage transmission
B. A computer control system that continuously monitors the state of input devices and makes decisions based on a custom program to control the state of output devices
C. A device used to increase the frequency of electrical signals
D. A device used to stabilize voltage in high-power applications

Q258: What is the function of a choke in electrical circuits?

A. To block or limit the flow of high frequency noise
B. To store electrical energy
C. To convert DC to AC
D. To amplify electrical signals

Answers

Answer 207: C. To interrupt the electrical circuit during an overload or short circuit

Explanation: A circuit breaker is designed to protect an electrical circuit from damage caused by excess current from an overload or short circuit. It automatically interrupts current flow after a fault is detected.

Answer 208: C. Ammeter

Explanation: An ammeter is specifically used to measure the flow of electrical current in a circuit.

Answer 209: B. To provide high current from a low current signal

Explanation: A relay is used in circuits to control a larger current circuit by a smaller current signal, often used for switching on/off larger loads with a smaller control signal.

Answer 210: B. Selector switch

Explanation: A selector switch is commonly used in industrial control systems to change the direction of motor rotation among other functions.

Answer 211: B. Generator

Explanation: A generator converts mechanical energy into electrical energy, typically by moving a conductor through a magnetic field to induce voltage.

Answer 212: B. To protect the circuit by breaking the connection when current exceeds a certain limit

Explanation: A fuse is a safety device that interrupts the power supply to a circuit when the current exceeds safe levels, preventing overheating and potential fires.

Answer 213: C. Stepper motor

Explanation: Stepper motors are used where precise speed control and positioning are necessary, as they move in discrete steps, allowing precise control.

Answer 214: C. Capacitor

Explanation: A capacitor stores electric charge and energy in an electric field, commonly used in electronic circuits to block direct current while allowing alternating current to pass, among other uses.

Answer 215: B. To protect against electrical shock by providing a path for fault current

Explanation: The primary purpose of the ground wire is to protect users from electrical shock by providing a safe path for fault current to travel back to the ground.

Answer 216: A. Electromagnetic induction

Explanation: Transformers operate on the principle of electromagnetic induction, where a change in magnetic flux induces a voltage across a coil.

Answer 217: C. Starter

Explanation: Starters are devices used to start and stop electric motors, providing the necessary current for starting the motor.

Answer 218: C. Proximity switch

Explanation: A proximity switch is used to detect the presence or absence of an object without physical contact, often utilized in industrial and automation settings.

Answer 219: C. By generating a voltage when two dissimilar metals are heated at one end

Explanation: A thermocouple works by generating a voltage difference between two dissimilar metals that are joined at one end when that junction is heated.

Answer 220: B. To adjust the speed of AC motors by varying the frequency of the electrical supply

Explanation: Variable Frequency Drives (VFDs) are used to control the speed of AC motors by varying the frequency and voltage of the electrical supply to the motor.

Answer 221: A. Surge protector

Explanation: A surge protector is used to protect electrical devices from voltage spikes by limiting the voltage supplied to an electric device by either blocking or shorting to ground any unwanted voltages above a safe threshold.

Answer 222: B. To convert alternating current (AC) to direct current (DC)

Explanation: A rectifier is an electrical device that converts AC, which periodically reverses direction, to DC, which flows in only one direction.

Answer 223: B. Ballast

Explanation: Ballasts regulate the current to the lamps and provide sufficient voltage to start fluorescent lamps.

Answer 224: B. To only allow current to flow in one direction

Explanation: A diode is a semiconductor device that allows current to flow in only one direction, effectively blocking the opposite direction to prevent reverse current.

Answer 225: C. To detect light levels and adjust lighting automatically

Explanation: Photocells are used in light-sensitive devices, commonly in outdoor lighting to turn lights on at dusk and off at dawn automatically.

Answer 226: B. Circuit breaker

Explanation: A circuit breaker is used to manually or automatically disconnect an electrical circuit in the event of an overload, providing a way to manually control the interruption of electrical flow.

Answer 227: B. To oppose changes in current

Explanation: Inductors oppose changes in current flowing through them due to the magnetic field created around the inductor when current flows.

Answer 228: B. Induction motor

Explanation: Induction motors are most commonly used in HVAC systems for fan control due to their durability and efficiency in variable load conditions.

Answer 229: A. By generating a voltage in response to mechanical stress

Explanation: Piezoelectric sensors work by generating an electrical charge in response to applied mechanical stress, commonly used in various sensing applications.

Answer 230: C. To convert electrical energy into linear motion

Explanation: Solenoids are devices that convert electrical energy into linear motion, commonly used to actuate mechanisms such as starting a car by turning the ignition key.

Answer 231: B. Twisted pair cable

Explanation: Twisted pair cables are commonly used in networking for transmitting both power (via PoE - Power over Ethernet) and data, catering to the needs of networked devices.

Answer 232: C. To dissipate heat to prevent overheating

Explanation: A heat sink in electronic devices is used to dissipate heat generated by electronic components, thereby preventing overheating and improving the reliability of the device.

Here are the answers and explanations for the questions regarding electrical equipment and their functions:

Answer 233: B. Ground Fault Circuit Interrupter (GFCI)

Explanation: A GFCI is designed to protect individuals from electric shock by detecting ground faults and interrupting the circuit.

Answer 234: C. To improve power factor in AC power systems

Explanation: Capacitor banks are used in electrical systems to improve the power factor, which helps to optimize the efficiency of the power system by reducing phase differences between voltage and current.

Answer 235: B. To distribute power across multiple circuits

Explanation: Busbars are used in electrical distribution systems to distribute electrical power across multiple circuits, providing a common connection point for incoming and outgoing currents.

Answer 236: A. Contactor

Explanation: Contactors are electrical devices used to control the flow of large electric currents using smaller control signals, commonly found in industrial and commercial electrical applications.

Answer 237: C. To provide a path for the magnetic flux

Explanation: In transformers, the core is crucial for providing a path for the magnetic flux to flow, which helps in the efficient transformation of electrical energy from one circuit to another.

Answer 238: C. Electrical properties like voltage, current, and resistance

Explanation: A multimeter is a versatile tool used primarily to measure various electrical properties, including voltage, current, and resistance.

Answer 239: B. Timer

Explanation: Timers are components used in circuits to delay the flow of electricity for a predetermined period, allowing for timed control of devices or processes.

Answer 240: B. To maintain constant voltage levels despite fluctuations in supply

Explanation: Stabilizers are used to maintain constant voltage levels in electrical systems, compensating for fluctuations in the input supply to provide a stable output.

Answer 241: B. Transformer

Explanation: Transformers are used to step down high voltages from power lines to usable voltage levels suitable for homes and businesses, ensuring safe and efficient power distribution.

Answer 242: C. To measure magnetic fields and convert them into a digital value

Explanation: Hall effect sensors are used in electronic devices to measure the intensity of magnetic fields and convert these measurements into digital values for further processing.

Answer 243: C. Three-way switch

Explanation: Three-way switches are commonly used in staircases and hallways to control lights from two different locations, allowing for convenient switching on and off from either end.

Answer 244: B. By measuring the time it takes for a sound wave to return to the sensor

Explanation: Ultrasonic sensors measure distance by emitting sound waves and calculating the time it takes for the echo to return, providing a method to determine the distance to an object.

Answer 245: A. To isolate part of a circuit while it is being worked on

Explanation: Electrical isolators are used to physically separate part of a circuit, ensuring that it is completely de-energized for maintenance or repair work, enhancing safety.

Answer 246: B. Lead-acid battery

Explanation: Lead-acid batteries are commonly used in uninterruptible power supply (UPS) systems due to their capacity to deliver high surge currents and reliable performance.

Answer 247: B. To control the brightness of a light

Explanation: Rheostats are adjustable resistors used in circuits to control the brightness of lights or the speed of motors by varying resistance.

Answer 248: A. Charge controller

Explanation: Charge controllers are used in battery-powered systems to prevent overcharging, managing the amount of power going into the battery to extend its life and maintain efficiency.

Answer 249: C. Thermostat

Explanation: Bimetallic strips are commonly found in thermostats, where they convert temperature changes into mechanical displacement to control heating or cooling systems.

Answer 250: B. Converts mechanical motion into an electrical signal to determine position

Explanation: Encoders are used in industrial automation to convert the position or motion of a shaft or axle to an analog or digital signal, allowing precise control of mechanical movement.

Answer 251: B. To protect against voltage spikes

Explanation: Varistors are used to protect circuits against high-voltage spikes. They change resistance to limit voltage to acceptable levels.

Answer 252: A. Photocell

Explanation: Photocells are light-sensitive devices used to control lighting and machinery automatically based on ambient light levels, commonly found in street lighting and automatic lighting systems.

Answer 253: A. Inverter

Explanation: Inverters are crucial in solar power systems, where they convert DC (direct current) from solar panels into AC (alternating current), suitable for home appliances and grid connection.

Answer 254: A. It operates at a constant speed up to its full load

Explanation: Synchronous motors operate at a constant speed synchronized to the frequency of the AC power supply, unlike asynchronous motors, which have a speed that can vary with the load.

Answer 255: C. In electronic drum pads to convert drum strikes into electrical signals

Explanation: Piezoelectric transducers are used in electronic drum pads, where they convert the force of drum strikes into electrical signals that can be processed into sound.

Answer 256: A. Wattmeter

Explanation: Wattmeters are specifically designed to measure and

Answer 257: B. A computer control system that continuously monitors the state of input devices and makes decisions based on a custom program to control the state of output devices

Explanation: Programmable Logic Controllers (PLCs) are used in industrial applications to manage the operations of machinery by monitoring inputs and controlling outputs according to a user-programmed logic.

Answer 258: A. To block or limit the flow of high frequency noise

Explanation: Chokes, or inductors, are used in circuits to block high-frequency alternating current (AC) in an electrical circuit, while allowing DC or low-frequency current to pass through, helping to suppress electrical noise.

Safety and First Aid

Q283: What should be used to treat a minor electrical burn?

A. Butter or oil-based ointment
B. Cool, clean water and cover with a sterile bandage
C. A tight bandage to prevent swelling
D. Heat to sterilize the area

Q284: What's the most critical step to take before performing first aid on a victim of electrical shock?

A. Check for responsiveness
B. Ensure the electrical source is disconnected
C. Apply cold compresses to burns
D. Move the victim to a different location

Q285: What kind of barrier should be used when performing CPR on an electrical shock victim?

A. Plastic wrap
B. A CPR mask or shield
C. A cloth towel
D. No barrier is necessary

Q286: What should be immediately done if someone has fallen from a height after receiving an electric shock?

A. Try to move them to see if they are conscious
B. Keep them still and stabilize the neck and back
C. Offer them water
D. Lift them up to a seated position

Q287: Which type of injuries are common in electricians due to the nature of their job?

A. Thermal burns
B. Electrical burns
C. Fractures and dislocations
D. All of the above

Q288: What is the main risk factor for developing arc eye or welder's flash from electrical work?

A. Working under direct sunlight
B. Exposure to bright UV light from welding arcs
C. Working at high altitudes
D. Long hours of focus on detailed work

Q289: How can bystanders assist immediately after an electrical accident?

A. By crowding around the victim to offer help
B. By using their phones to record helpful information
C. By maintaining a safe distance and directing others to stay back
D. By offering food and drink to the victim

Q290: What are the key elements of an effective electrical safety training program?

A. Practical demonstrations only
B. Classroom lectures only
C. A combination of theory, practical demonstrations, and regular drills
D. Reading and passing a test on safety protocols

Q291: What does the term "dielectric" refer to in the context of personal protective equipment (PPE)?

A. PPE that is biodegradable
B. PPE designed to absorb electric shock
C. PPE made from natural materials
D. PPE that is non-conductive and provides insulation from electrical hazards

Q292: What should be done after using a fire extinguisher on an electrical fire?

A. Return the extinguisher to its original location
B. Leave the extinguisher at the scene for professional evaluation
C. Have the extinguisher refilled or replaced immediately
D. Dispose of the extinguisher

Q293: What is the importance of knowing the location of circuit breakers in a workplace?

A. To monitor energy consumption
B. To use it as a storage space
C. To quickly deactivate power in an emergency
D. For routine cleaning

Q294: What is one of the first symptoms of overexposure to electrical discharge?

A. Increased appetite
B. Tingling or numbness
C. A rash on the skin
D. Drowsiness

Q295: What routine should be followed for the maintenance of electrical tools and equipment?

A. Monthly professional inspections
B. Only service when a malfunction occurs
C. Routine checks and servicing as per manufacturer's guidelines
D. Annual replacement, regardless of condition

Q296: Why should jewelry be removed before working with electricity?

130

A. It can conduct electricity, increasing the risk of electric shock
B. It might get dirty
C. It can reflect light and dazzle colleagues
D. It is generally considered unprofessional

Q297: What should be included in an electrical workplace's first aid kit specifically for electrical injuries?

A. Burn creams and ointments
B. Insulating gloves
C. Pain relievers
D. All of the above

Q298: What is a safe distance to maintain from electrical panels to ensure safety?

A. At least 3 feet
B. At least 1 foot
C. No specific distance, as long as the panel is closed
D. At least 5 feet

Q299: How should one approach a victim still in contact with an electrical source?

A. With wet hands to cool them down
B. With metal tools to free them from the source
C. Using non-conductive materials to separate them from the source
D. None of the above; ensure the power source is off first

Q300: What are the signs that electrical equipment may be faulty and pose a safety risk?

A. Occasional sparks during operation
B. A slight burning smell
C. Unusual noises during use
D. All of the above

Q301: How often should safety drills be conducted in an environment with high electrical risks?

A. Annually
B. Bi-annually
C. Quarterly
D. Monthly

Q302: What should be clearly marked on all electrical panels and circuits for safety?

A. The name of the electrician
B. Warning signs and labels indicating the potential hazard
C. Decorative markings for easy identification
D. The installation date

Q303: What procedure should be followed if an electrical tool malfunctions while being used?

A. Continue using it until the task is completed
B. Turn off the tool, unplug it, and tag it out of service
C. Try to repair it immediately
D. Ignore the malfunction if it seems minor

Q304: What role does hydration play in electrical safety?

A. Dehydrated individuals are more prone to static discharge
B. Hydration levels do not affect electrical safety
C. Well-hydrated individuals are less likely to experience severe shocks
D. None; hydration only affects physical comfort

Q305: What is the standard procedure for updating safety protocols in an electrical facility?

A. Whenever a new manager is appointed
B. After any electrical incident
C. As new technology is introduced
D. Annually, regardless of other factors

Q306: Why is it critical to have a clear path to exits in areas with high electrical risk?

A. To comply with health and beauty standards
B. To facilitate quick evacuation in case of an emergency
C. To make room for more equipment
D. To allow for easier cleaning

Q307: What specific training should be provided for workers who are at risk of arc flash exposure?

A. Basic CPR only
B. Advanced chemical handling
C. Specific arc flash hazard awareness and how to minimize risks
D. General safety videos annually

Q308: How should electrical injuries be documented in the workplace?

A. Informally, between employees
B. Not necessary unless severe
C. In a dedicated incident log as part of an ongoing safety record
D. Only through verbal reports

Q309: What is the recommended treatment for someone who has been shocked and is showing signs of cardiac arrest?

A. Immediately start CPR and use an AED if available.
B. Give them water to drink.
C. Wait for them to regain consciousness on their own.
D. Move them to a cooler location.

Q310: What is the best practice for storing electrical power tools to ensure safety?

A. Keep them plugged in but switched off.
B. Store them in a damp location.
C. Ensure they are clean and stored in a dry, secure place.
D. Leave them on the ground for easy access.

Q311: What type of gloves should be worn to handle live electrical wires?

A. Leather gloves.
B. Rubber insulating gloves.
C. Fabric gloves.
D. No gloves are needed if the wires are not frayed.

Q312: What immediate steps should be taken if an electrical panel catches fire?

A. Try to extinguish the fire with water.
B. Immediately shut off the main power, if it's safe to do so, and use a CO2 fire extinguisher.
C. Open the panel to let the smoke out.
D. Remove all nearby materials that are flammable.

Q313: How should one act if experiencing a tingling sensation while using electrical equipment?

A. Stop using the equipment immediately, disconnect it, and report the issue.
B. Ignore the sensation as it is a common occurrence.
C. Check if others are experiencing the same sensation before stopping work.
D. Increase the speed of work to finish before any incident occurs.

Q314: What is the first action to take if you notice someone being electrocuted?

A. Rush to pull them away from the source.
B. Use a metal rod to disconnect them.
C. Shut off the power source immediately, if possible.
D. Immediately touch them to ground the electricity.

Q315: Why is it important to install GFCI (Ground Fault Circuit Interrupter) devices in wet areas?

A. They prevent overloads.
B. They interrupt the circuit if a ground fault is detected, preventing potential shocks.
C. They make the circuit work faster.
D. They reduce energy consumption.

Q316: How should you approach the rescue of someone in contact with high voltage?

A. With bare hands to ensure quick action.
B. Using insulated tools and ensuring you are properly insulated.
C. Call for help but do not approach the person.
D. Spray water on the person to reduce the voltage.

Q317: What are the symptoms of acute electrical burns?

A. Nausea and headache.
B. Charred or blackened skin at the contact points.
C. Increased energy.
D. A feeling of warmth without any visible marks.

Q318: What should be the content of safety training for new employees working with electricity?

A. The history of the company.
B. Basic duties and minimal safety information.
C. Comprehensive electrical safety, emergency procedures, and hands-on practice.
D. Only the theoretical aspects of electricity.

Q319: How can one ensure that an electrical workspace is safe from tripping hazards?

A. Regularly inspect areas for loose wires and cords and secure them.
B. Keep all wires and cords in visible areas.
C. Use more wireless devices to reduce the number of cords.
D. Ignore small wires as they do not pose serious risks.

Q320: What is the significance of having a detailed and accessible electrical schematic in the workplace?

A. It is useful for new equipment installation.

134

B. It helps in quick troubleshooting and safe handling of emergencies.
C. It is only necessary for large installations.
D. It provides general information about the building's layout.

Q321: What precautions should be taken when using extension cords in a workplace?

A. Run them under carpets to avoid exposure.
B. Use them permanently instead of installing more outlets.
C. Check for the correct rating for the job and ensure they are not daisy-chained or overloaded.
D. Use as many as needed regardless of their power ratings.

Q322: How should flammable materials be stored in relation to electrical equipment?

A. Directly beside the equipment for easy access.
B. In a separate, designated non-flammable area.
C. On top of electrical equipment to save space.
D. Within a few feet to keep all materials together.

Q323: What regular maintenance should be performed on electrical tools to ensure safety?

A. Occasional cleaning with water.
B. Regular checking for wear and tear, testing for functionality, and ensuring safety features are operational.
C. Only replace parts when they break.
D. Lubrication with oils to enhance conductivity.

Q324: What emergency equipment should be readily available in all electrical maintenance areas?

A. A bucket of sand.
B. Fire extinguishers suitable for electrical fires and a basic first aid kit.
C. A water hose.
D. Additional electrical tools.

Q325: Why is it important to clearly label all circuit breakers and fuses?

A. To enhance aesthetic appeal.
B. To ensure that the correct circuits can be quickly and safely shut down or activated.
C. To comply with decorating guidelines.
D. Labeling has no real safety benefit.

Q326: How should injuries from electrical work be reported and recorded?

A. In an informal conversation.
B. Through a formal process that includes documenting the incident in detail and reporting to the appropriate management.
C. Only record if the injury is serious.
D. Reporting is optional based on the worker's discretion.

Q327: What special considerations should be given to electrical safety during adverse weather conditions?

A. Continue normal operations as electricity is not affected by weather
B. Suspend all outdoor or risky electrical operations until conditions improve.
C. Only allow senior staff to work with electricity.
D. Increase the number of workers to speed up work before the weather worsens.

Q328: What routine inspections should be conducted on electrical installations in commercial buildings?

A. Daily inspections by all staff.
B. Weekly inspections by maintenance personnel.
C. Periodic detailed inspections by qualified electricians to ensure compliance with safety standards.
D. Inspections only when visible problems occur.

Q329: How should electric shock incidents be analyzed?

A. As isolated, unpredictable events.
B. Through a detailed incident analysis process to prevent future occurrences.
C. Quickly and without in-depth investigation.
D. By untrained personnel to ensure impartiality.

Q330: What kind of training should be provided to employees to handle electrical emergencies?

A. Basic overview of electrical systems.
B. Detailed, regularly updated training including emergency response and first aid.
C. Only legally required minimum training.
D. Training is not necessary if safety measures are in place.

Q331: What is the best practice for managing cables on the floor of an office or worksite?

A. Use cable covers or guards to prevent tripping and protect the cables.
B. Allow cables to lay freely for easy access.
C. Tape cables down with duct tape.
D. Color code cables for identification without physical management.

Q332: How should one verify that an electrical tool or appliance is de-energized before performing maintenance?

A. Guess based on the last usage.
B. Check with a multimeter or voltage tester.
C. Ask a colleague if the tool is off.
D. Assume it is de-energized if it's not operating.

Answers

Answer 283: B. Cool, clean water and cover with a sterile bandage

Explanation: For minor electrical burns, the recommended treatment is to cool the burn under running water for at least ten minutes and then cover it with a sterile bandage to protect the area.

Answer 284: B. Ensure the electrical source is disconnected

Explanation: Before administering first aid to a victim of electrical shock, the most critical step is to ensure the electrical source is disconnected to prevent further injury and ensure safety for both the victim and the rescuer.

Answer 285: B. A CPR mask or shield

Explanation: When performing CPR on an electrical shock victim, using a CPR mask or shield is essential to prevent direct contact with the victim's bodily fluids, thereby protecting both the rescuer and the victim.

Answer 286: B. Keep them still and stabilize the neck and back

Explanation: If someone has fallen from a height after receiving an electric shock, it is crucial to keep them still and stabilize their neck and back to prevent further injury, particularly to the spinal cord.

Answer 287: D. All of the above

Explanation: Electricians are commonly at risk for various injuries including thermal burns from heat or fire, electrical burns from direct contact with current, and fractures or dislocations from falls or other accidents.

Answer 288: B. Exposure to bright UV light from welding arcs

Explanation: The main risk factor for developing arc eye or welder's flash is exposure to intense UV light emitted by welding arcs, which can burn the cornea and cause significant eye discomfort.

Answer 289: C. By maintaining a safe distance and directing others to stay back

Explanation: After an electrical accident, bystanders can assist by maintaining a safe distance from the accident scene and directing others to stay back, helping to secure the area and prevent additional injuries.

Answer 290: C. A combination of theory, practical demonstrations, and regular drills

Explanation: Effective electrical safety training should include a combination of theoretical learning, practical demonstrations, and regular emergency response drills to ensure comprehensive knowledge and preparedness.

Answer 291: D. PPE that is non-conductive and provides insulation from electrical hazards

Explanation: Dielectric PPE refers to personal protective equipment that is non-conductive and provides insulation to protect the wearer from electrical hazards.

Answer 292: C. Have the extinguisher refilled or replaced immediately

Explanation: After using a fire extinguisher on an electrical fire, it is crucial to have the extinguisher refilled or replaced immediately to ensure it is ready for use in case another emergency arises.

Answer 293: C. To quickly deactivate power in an emergency

Explanation: Knowing the location of circuit breakers in a workplace is important as it allows personnel to quickly deactivate power sources in an emergency, preventing further hazards and facilitating safer emergency responses.

Answer 294: B. Tingling or numbness

Explanation: One of the first symptoms of overexposure to electrical discharge can be tingling or numbness, which indicates potential nerve damage from the electric current.

Answer 295: C. Routine checks and servicing as per manufacturer's guidelines

Explanation: Routine maintenance checks and servicing of electrical tools and equipment as per the manufacturer's guidelines are essential to ensure they remain in good working condition and safe to use.

Answer 296: A. It can conduct electricity, increasing the risk of electric shock

Explanation: Jewelry should be removed before working with electricity because it can conduct electricity, potentially leading to an increased risk of electric shock.

Answer 297: D. All of the above

Explanation: An electrical workplace's first aid kit should include burn creams and ointments for treating burns, insulating gloves to handle electrical components safely, and pain relievers to manage discomfort from injuries.

Answer 298: A. At least 3 feet

Explanation: Maintaining at least 3 feet of clearance from electrical panels is recommended to ensure safety and provide enough space for operating and servicing the panel safely.

Answer 299: D. None of the above; ensure the power source is off first

Explanation: When approaching a victim still in contact with an electrical source, do not attempt to touch or move them until you are certain the power source is turned off to avoid conducting electricity through your body.

Answer 300: D. All of the above

Explanation: Signs that electrical equipment may be faulty and pose a safety risk include occasional sparks during operation, a slight burning smell, or unusual noises during use.

Answer 301: C. Quarterly

Explanation: In environments with high electrical risks, safety drills should be conducted quarterly to ensure all personnel are familiar with emergency procedures and can respond effectively in case of an incident.

Answer 302: B. Warning signs and labels indicating the potential hazard

Explanation: All electrical panels and circuits should be clearly marked with warning signs and labels indicating the potential hazard to inform and protect workers and maintenance personnel.

Answer 303: B. Turn off the tool, unplug it, and tag it out of service

Explanation: If an electrical tool malfunctions while being used, immediately turn it off, unplug it, and tag it as out of service to prevent others from using it until it has been checked and repaired.

Answer 304: A. Dehydrated individuals are more prone to static discharge

Explanation: Hydration plays a role in electrical safety as dehydrated individuals are more prone to static discharge, which can be dangerous in environments with high electrical risks.

Answer 305: B. After any electrical incident

Explanation: Safety protocols in an electrical facility should be updated after any electrical incident to address any discovered vulnerabilities and improve safety measures based on recent findings.

Answer 306: B. To facilitate quick evacuation in case of an emergency

Explanation: Keeping a clear path to exits in areas with high electrical risk is critical to facilitate quick and safe evacuation in case of an emergency, helping to prevent injuries during rapid exits.

Answer 307: C. Specific arc flash hazard awareness and how to minimize risks

Explanation: Workers at risk of arc flash exposure should receive specific training on arc flash hazards, including awareness and techniques to minimize risks, to ensure they understand the dangers and how to protect themselves effectively.

Answer 308: C. In a dedicated incident log as part of an ongoing safety record

Explanation: Electrical injuries and incidents should be documented in a dedicated incident log within the workplace to maintain an ongoing safety record, which helps in assessing risk factors and improving safety protocols.

Answer 309: A. Immediately start CPR and use an AED if available.

Explanation: For someone who has been shocked and shows signs of cardiac arrest, it is crucial to start CPR immediately and use an Automated External Defibrillator (AED) if available, to attempt to restore a normal heart rhythm.

Answer 310: C. Ensure they are clean and stored in a dry, secure place.

Explanation: Storing electrical power tools in a clean, dry, and secure place is essential to ensure they remain in good working condition and are safe to use. This prevents damage from moisture and unauthorized access.

Answer 311: B. Rubber insulating gloves.

Explanation: Rubber insulating gloves are essential for handling live electrical wires as they provide insulation against electrical shock, protecting the user from potential harm.

Answer 312: B. Immediately shut off the main power, if it's safe to do so, and use a CO2 fire extinguisher.

Explanation: If an electrical panel catches fire, it's crucial to shut off the main power if possible and use a CO2 fire extinguisher, which is effective on electrical fires without conducting electricity.

Answer 313: A. Stop using the equipment immediately, disconnect it, and report the issue.

Explanation: Experiencing a tingling sensation while using electrical equipment could indicate a potential electrical fault. It's important to stop using the equipment immediately, disconnect it, and report the issue to prevent further hazards.

Answer 314: C. Shut off the power source immediately, if possible.

Explanation: If you notice someone being electrocuted, the first step is to shut off the power source immediately, if possible, to stop the flow of electricity and prevent further injury.

Answer 315: B. They interrupt the circuit if a ground fault is detected, preventing potential shocks.

Explanation: GFCI devices are important in wet areas because they can detect ground faults and quickly interrupt the circuit, significantly reducing the risk of electric shock.

Answer 316: B. Using insulated tools and ensuring you are properly insulated.

Explanation: When approaching a rescue involving high voltage, it is crucial to use insulated tools and ensure you are properly insulated to prevent personal injury from electrical shock.

Answer 317: B. Charred or blackened skin at the contact points.

Explanation: Acute electrical burns often result in charred or blackened skin at the contact points, which are the visible signs of severe electrical burns.

Answer 318: C. Comprehensive electrical safety, emergency procedures, and hands-on practice.

Explanation: Safety training for new employees working with electricity should be comprehensive, including details on electrical safety, emergency procedures, and hands-on practice to ensure they are well-prepared for their roles.

Answer 319: A. Regularly inspect areas for loose wires and cords and secure them.

Explanation: To ensure an electrical workspace is safe from tripping hazards, it's important to regularly inspect the area for loose wires and cords and secure them properly to prevent accidents.

Answer 320: B. It helps in quick troubleshooting and safe handling of emergencies.

Explanation: Having a detailed and accessible electrical schematic in the workplace is crucial for quick troubleshooting and safe handling of emergencies by providing clear information on how the electrical system is configured.

Answer 321: C. Check for the correct rating for the job and ensure they are not daisy-chained or overloaded.

Explanation: When using extension cords, it's important to check that they are correctly rated for the job at hand and ensure they are not daisy-chained or overloaded to prevent electrical hazards.

Answer 322: B. In a separate, designated non-flammable area.

Explanation: Flammable materials should be stored in a separate, designated non-flammable area away from electrical equipment to reduce the risk of fire and ensure safety in the workplace.

Certainly! Here are the responses formatted as requested:

Answer 323: B. Regular checking for wear and tear, testing for functionality, and ensuring safety features are operational.

Explanation: Regular maintenance of electrical tools should include checking for wear and tear, testing for functionality, and ensuring that all safety features are operational. This practice helps prevent accidents and ensures the tools are safe to use.

Answer 324: B. Fire extinguishers suitable for electrical fires and a basic first aid kit.

Explanation: Emergency equipment in electrical maintenance areas should include fire extinguishers that are suitable for electrical fires (such as CO2 or dry chemical extinguishers) and a basic first aid kit. This equipment is essential for handling potential injuries and fires promptly and safely.

Answer 325: B. To ensure that the correct circuits can be quickly and safely shut down or activated.

Explanation: Clearly labeling all circuit breakers and fuses is crucial. It ensures that during maintenance or an emergency, the correct circuits can be quickly and safely shut down or activated, thus reducing the risk of injury and damage.

Answer 326: B. Through a formal process that includes documenting the incident in detail and reporting to the appropriate management.

Explanation: Injuries from electrical work should be reported and recorded through a formal process, including documenting the incident in detail and reporting to appropriate management. This ensures proper follow-up and preventive measures are implemented.

Answer 327: B. Suspend all outdoor or risky electrical operations until conditions improve.

Explanation: During adverse weather conditions, it is important to suspend all outdoor or risky electrical operations until conditions improve to prevent accidents caused by wet conditions, lightning, or other weather-related factors.

Answer 328: C. Periodic detailed inspections by qualified electricians to ensure compliance with safety standards.

Explanation: Electrical installations in commercial buildings should be subjected to periodic detailed inspections by qualified electricians. This ensures they comply with safety standards and helps identify and rectify potential hazards.

Answer 329: B. Through a detailed incident analysis process to prevent future occurrences.

Explanation: Electric shock incidents should be analyzed through a detailed incident analysis process. This helps understand what went wrong and implement measures to prevent future occurrences.

Answer 330: B. Detailed, regularly updated training including emergency response and first aid.

Explanation: Employees should receive detailed and regularly updated training that includes emergency response and first aid. This ensures they are equipped to handle electrical emergencies effectively and safely.

Answer 331: A. Use cable covers or guards to prevent tripping and protect the cables.

Explanation: The best practice for managing cables on the floor of an office or worksite is to use cable covers or guards. This prevents tripping hazards and protects the cables from damage, ensuring a safer and more organized workspace.

Motor Controls and Automation Systems

Q333: What is the primary function of a motor starter in an automation system?

A. To reduce the motor's operating speed.
B. To safely start and stop the motor by controlling the application of power.
C. To monitor the temperature of the motor.
D. To increase the voltage supplied to the motor.

Q334: Which device is typically used to change the direction of a motor?

A. Contactor
B. Inverter
C. Reversing starter
D. Solenoid

Q335: What is the role of a Variable Frequency Drive (VFD) in motor control?

A. To maintain a constant voltage to the motor.
B. To adjust the motor speed by varying the frequency of the power supplied to the motor.
C. To convert AC power to DC power.
D. To protect the motor from overloads.

Q336: In what type of control system is PLC (Programmable Logic Controller) commonly used?

A. Basic manual control systems.
B. Advanced automated control systems.
C. Low-voltage lighting systems.
D. High-power transmission systems.

Q337: What is the purpose of using an encoder in motor control systems?

A. To convert electrical noise to usable signals.
B. To regulate voltage.
C. To detect and communicate the motor's speed and position.
D. To insulate the motor circuits.

Q338: Which component is crucial for protecting a motor from overcurrent conditions?

A. Fuse
B. Capacitor
C. Overload relay
D. Transformer

Q339: How does a soft starter differ from a traditional motor starter?

A. It starts motors at a lower voltage.
B. It gradually increases the voltage to the motor to reduce mechanical stress and electrical peaking.
C. It uses a series of relays for gradual startup.
D. It provides a constant voltage from start to finish.

Q340: What is a servo motor primarily used for in automation systems?

A. High power applications.
B. Precise control of angular or linear position, velocity, and acceleration.
C. General switching operations.
D. Voltage regulation.

Q341: What type of motor control is typically used in conveyor systems?

A. Servo control
B. Stepper control
C. AC motor drive control
D. Manual switch control

Q342: Which sensor is commonly used to measure the speed of a motor shaft?

A. Thermocouple
B. Proximity sensor
C. Tachometer
D. Photocell

Q343: What does a Motor Control Center (MCC) house?

A. Only fuses and isolators.
B. Multiple motor starters, drives, and associated control equipment.
C. Only transformers and rectifiers.
D. Lighting controls only.

Q344: How does an interlock function in a motor control circuit?

A. To disconnect the power automatically in case of an emergency.
B. To prevent conflicting operations that could cause damage.
C. To boost the signal strength to the motor.
D. To reduce the operational speed of all connected motors.

Q345: What is the main advantage of using DC motors in automation systems compared to AC motors?

A. Higher power output.
B. Simpler construction.
C. Easier speed and torque control.
D. More energy-efficient at high speeds.

Q346: What purpose does a brake serve in an electric motor system?

A. To increase the motor's speed.
B. To ensure the motor runs continuously.
C. To safely slow or stop the motor.
D. To convert the motor's kinetic energy into electrical energy.

Q347: What is a common use for proximity sensors in automated systems?

A. To measure the temperature of the system.
B. To detect the presence or absence of objects near the sensor.
C. To control the voltage supplied to the motor.
D. To store data about the system's operations.

Q348: What kind of switch is commonly used to control the operation of a large industrial motor?

A. Toggle switch.
B. Limit switch.
C. Push button switch.
D. Mercury switch.

Q349: How are actuators used in automation systems?

A. To manually control valves.
B. To provide real-time data logging.
C. To convert electrical energy into mechanical motion.
D. To insulate high-voltage circuits.

Q350: What is a main function of human-machine interface (HMI) systems in motor control?

A. To cool down the motors.
B. To provide a user-friendly platform for monitoring and controlling the machines.
C. To mechanically adjust machine components.
D. To enhance the electrical conductivity between machine components.

Q351: What does a phase failure relay protect against?

146

A. Voltage overload.
B. Loss of one or more phases in a three-phase system, preventing potential motor damage.
C. Temperature fluctuations.
D. Mechanical wear and tear.

Q352: How does an overload relay protect an electric motor?

A. By shutting down the motor when it overheats.
B. By speeding up the motor when a load is detected.
C. By providing additional voltage during high-load conditions.
D. By cooling the motor automatically.

Q353: What type of motor is best suited for applications requiring frequent start and stop actions?

A. Single-phase AC motor.
B. Three-phase AC motor.
C. Stepper motor.
D. Synchronous motor.

Q354: In what application might you use a stepper motor?

A. For high-speed, high-torque operations.
B. In precision positioning systems like 3D printers or robotic arms.
C. For heavy industrial pumping.
D. In high-voltage transmission.

Q355: What role does a contactor play in motor control?

A. It provides a physical barrier for safety.
B. It adjusts the motor's operational frequency.
C. It serves as a heavy-duty relay to control the power delivered to the motor.
D. It measures the electrical output of the motor.

Q356: What is the primary consideration when designing a motor control system for a hazardous environment?

A. Ensuring the system can withstand extreme temperatures.
B. The system's color coding.
C. The robustness of security features.
D. Using components that are explosion-proof and sealed against contaminants.

Q357: How does a frequency inverter benefit a motor control system?

A. It provides a fixed frequency to ensure stable motor speed.
B. It varies the frequency of the supply to the motor, allowing variable speed control.
C. It converts the motor's output frequency to a higher level for industrial applications.
D. It stabilizes the frequency to prevent fluctuations in power supply.

Q358: What kind of protection does a thermal overload relay offer to a motor?

A. It protects against short circuits.
B. It protects against power surges.
C. It protects the motor from overheating due to excessive current.
D. It insulates the motor from environmental heat sources.

Q359: What is the purpose of a motor soft starter in an industrial setting?

A. To reduce the mechanical stress on motor and driven equipment during startup.
B. To increase the energy consumption during startup.
C. To decrease the efficiency of the motor.
D. To shut down the motor quickly during an emergency.

Q360: Which device is essential for reversing the direction of a DC motor in an automated system?

A. Variable Frequency Drive (VFD)
B. Reversing contactor
C. Solid-state relay
D. Capacitor

Q361: How does an Automatic Transfer Switch (ATS) function in motor control systems?

A. By converting AC to DC.
B. By manually switching the power source when necessary.
C. By automatically switching the power source to maintain continuous operation.
D. By decreasing the power supply to motors.

Q362: What is the function of limit switches in automation systems?

A. To control the speed of motors.
B. To measure the current flowing through the system.
C. To detect the physical movement or position of an object.

D. To convert mechanical motion into electrical signals.

Q363: How is torque control implemented in motor control systems?

A. Through manual adjustment only.
B. Using hydraulic systems.
C. Via electronic control, such as with a VFD that adjusts the power supplied to the motor.
D. Through the use of gears and levers.

Q364: What safety feature is integral to modern motor starters to protect against electrical faults?

A. Optical sensors.
B. Built-in circuit breakers.
C. Noise filters.
D. Heat sinks.

Q365: What role do programmable logic controllers (PLCs) play in automated motor control systems?

A. They provide backup power to systems.
B. They primarily increase motor speed.
C. They control the sequence of operations and automate processes.
D. They decrease the automation complexity.

Q366: What is the main advantage of using an encoder with a servo motor?

A. It decreases the motor's power consumption.
B. It provides precise control over the motor's position, speed, and acceleration.
C. It simplifies the motor's electrical connections.
D. It enhances the motor's torque.

Q367: Why is grounding important in motor control panels?

A. To improve the panel's operational speed.
B. To prevent static electricity buildup.
C. To ensure safety by dissipating fault currents.
D. To enhance the aesthetic appeal of the panel.

Q368: What is the role of a human-machine interface (HMI) in an industrial automation system?

A. To manually control the machines without any feedback.

B. To display process data and facilitate user control over the system.
C. To restrict access to machine controls.
D. To provide lighting for the industrial area.

Q369: How do thermal protectors in motors help prevent damage?

A. By cooling the motor with a fan.
B. By detecting excessive heat and interrupting the power supply.
C. By lubricating the motor.
D. By increasing the voltage supply during high-temperature conditions.

Q370: What is the principle of operation of a magnetic contactor in motor control?

A. It uses a magnetic field to mechanically switch the motor circuit on or off.
B. It creates a permanent magnetic bond for security.
C. It uses magnetism to increase motor speed.
D. It relies on magnetic attraction to reduce power usage.

Q371: What types of sensors are typically used in motor control systems to ensure efficiency and safety?

A. Light sensors only.
B. Temperature and vibration sensors.
C. Sound sensors.
D. Humidity sensors.

Q372: What is an inverter used for in motor control applications?

A. To maintain a constant temperature.
B. To convert DC to AC, enabling precise speed control of AC motors.
C. To mechanically switch gears in the motor.
D. To reduce the weight of the motor.

Q373: How do phase monitors protect motors in industrial environments?

A. By balancing the load automatically.
B. By detecting phase loss, phase imbalance, and phase reversal to prevent motor damage.
C. By increasing the frequency of the electrical supply.
D. By monitoring the color phases used in wiring.

Q374: What is the function of a motor control center (MCC) in large industrial plants?

A. To serve as the main power distribution method.
B. To centralize control of various motors, providing power, control, and monitoring capabilities.
C. To monitor the environmental conditions around motors.
D. To house the plant's lighting controls.

Q375: What kind of motor is typically controlled with a direct online starter?

A. High-torque motors.
B. Low-power motors.
C. Small, single-phase motors.
D. Large, three-phase motors.

Q376: How are capacitors used in conjunction with electric motors?

A. To store extra electrical energy only.
B. To enhance the appearance of the motor.
C. To improve power factor and provide necessary phase shifts for starting torque.
D. To serve as a backup power source.

Q377: What considerations are taken into account when selecting a motor for an automated system?

A. Color and size of the motor.
B. Power, speed, torque, and the control system compatibility.
C. The brand popularity.
D. The noise level it produces only.

Q378: How is speed controlled in stepper motors?

A. By varying the voltage supplied to the motor.
B. By changing the frequency of the input pulses.
C. By using a fixed gear ratio.
D. By manual adjustment only.

Q379: What is the importance of using overload relays in motor control circuits?

A. They provide additional power to the motor.
B. They protect the motor from excessive current by breaking the circuit when necessary.
C. They make the motor run faster.
D. They cool the motor during operation.

Q380: In what way do circuit breakers in motor control panels enhance safety?

A. By decorating the panel.

- B. By interrupting power flow in case of an overload, preventing potential motor damage or electrical fires.
- C. By reducing the operational efficiency of the system.
- D. By increasing the voltage supply during peak hours.

Q381: How does a synchronous motor controller differ from an asynchronous motor controller?

- A. It allows for speed and position control that is in sync with the power grid frequency.
- B. It operates at variable frequencies only.
- C. It does not use electricity.
- D. It is used for small, portable motors only.

Q382: What are the benefits of integrating automation systems with motor controls?

- A. To reduce the need for human intervention, increase efficiency, and improve system reliability.
- B. To increase the energy consumption significantly.
- C. To complicate the operational procedures.
- D. To only allow manual control over the system.

Answers

Answer 333: B. To safely start and stop the motor by controlling the application of power.

Explanation: A motor starter's primary function is to safely manage the application of power to a motor, facilitating a controlled start and stop that protects both the motor and connected systems from abrupt electrical surges and mechanical stress.

Answer 334: C. Reversing starter.

Explanation: A reversing starter is typically used in motor control systems to change the direction of a motor. It allows the motor to run forward or backward by reversing the current flow within the system.

Answer 335: B. To adjust the motor speed by varying the frequency of the power supplied to the motor.

Explanation: A Variable Frequency Drive (VFD) controls the speed of an electric motor by varying the frequency of the electrical power supplied to the motor. This allows for precise speed control based on the requirements of the application.

Answer 336: B. Advanced automated control systems.

Explanation: Programmable Logic Controllers (PLCs) are extensively used in advanced automated control systems where flexibility, precision, and reliability are crucial. They can be programmed to perform a wide range of automation tasks effectively.

Answer 337: C. To detect and communicate the motor's speed and position.

Explanation: Encoders in motor control systems are used to detect and communicate the speed and position of the motor to the control system, enabling precise control over the motor's movements and operations.

Answer 338: C. Overload relay.

Explanation: An overload relay is crucial for protecting motors from overcurrent conditions. It detects excessive current flow and interrupts the power supply to prevent damage to the motor.

Answer 339: B. It gradually increases the voltage to the motor to reduce mechanical stress and electrical peaking.

Explanation: A soft starter differs from traditional starters by gradually increasing the voltage supplied to the motor. This reduces mechanical stress and electrical peaking during startup, thereby extending the motor's lifespan and improving energy efficiency.

Answer 340: B. Precise control of angular or linear position, velocity, and acceleration.

Explanation: Servo motors are used in automation systems for their ability to provide precise control of angular or linear position, velocity, and acceleration, essential for applications requiring exact movements, such as in robotics and CNC machinery.

Answer 341: C. AC motor drive control.

Explanation: AC motor drive control is typically used in conveyor systems due to its ability to control the speed and torque of the conveyor motor, ensuring efficient and adaptable operation.

Answer 342: C. Tachometer.

Explanation: A tachometer is commonly used to measure the speed of a motor shaft. It provides accurate readings of rotational speed, which is crucial for monitoring and control in various industrial applications.

Answer 343: B. Multiple motor starters, drives, and associated control equipment.

Explanation: A Motor Control Center (MCC) houses multiple motor starters, drives, and associated control equipment. It centralizes control and provides a structured means of managing various motors in a facility.

Answer 344: B. To prevent conflicting operations that could cause damage.

Explanation: An interlock in a motor control circuit functions to prevent conflicting operations, such as the simultaneous starting of incompatible functions, which could lead to mechanical damage or safety hazards.

Answer 345: C. Easier speed and torque control.

Explanation: DC motors are favored in automation systems for their easier speed and torque control compared to AC motors. This makes them suitable for applications requiring precise motion control.

Answer 346: C. To safely slow or stop the motor.

Explanation: The primary purpose of a brake in an electric motor system is to safely slow or stop the motor, ensuring controlled and safe operations, particularly in applications where precise stopping is crucial.

Answer 347: B. To detect the presence or absence of objects near the sensor.

Explanation: Proximity sensors are commonly used in automated systems to detect the presence or absence of objects. They are crucial for applications where contactless detection is required for operational safety and efficiency.

Answer 348: C. Push button switch.

Explanation: Push button switches are commonly used to control the operation of large industrial motors due to their ease of use and ability to provide quick activation or deactivation of the motor.

Answer 349: C. To convert electrical energy into mechanical motion.

Explanation: Actuators in automation systems are used to convert electrical energy into mechanical motion. This is essential for driving mechanical movements in automated equipment, such as opening valves or moving mechanical arms.

Answer 350: B. To provide a user-friendly platform for monitoring and controlling the machines.

Explanation: Human-machine interface (HMI) systems in motor control provide a user-friendly platform for operators to monitor and control machines. They facilitate interaction between the user and the machinery, allowing for efficient operation and monitoring.

Answer 351: B. Loss of one or more phases in a three-phase system, preventing potential motor damage.

Explanation: A phase failure relay protects against the loss of one or more phases in a three-phase system. This protection is crucial as phase loss can cause motors to operate inefficiently, leading to potential damage.

Answer 352: A. By shutting down the motor when it overheats.

Explanation: An overload relay protects an electric motor by automatically shutting it down when it overheats due to excessive current, preventing damage and potential fire hazards.

Answer 353: C. Stepper motor.

Explanation: Stepper motors are best suited for applications requiring frequent start and stop actions. They provide precise control over motion, making them ideal for tasks that require exact positioning.

Answer 354: B. In precision positioning systems like 3D printers or robotic arms.

Explanation: Stepper motors are used in precision positioning systems, such as 3D printers and robotic arms, due to their ability to precisely control movement in steps, allowing for exact positioning and movement.

Answer 355: C. It serves as a heavy-duty relay to control the power delivered to the motor.

Explanation: A contactor in motor control acts as a heavy-duty relay that controls the power delivered to the motor. It is capable of handling high currents, which is essential for starting and stopping large motors.

Answer 356: D. Using components that are explosion-proof and sealed against contaminants.

Explanation: When designing motor control systems for hazardous environments, it is critical to use components that are explosion-proof and sealed against contaminants to prevent accidents and ensure the safety of the system and its operators.

Answer 357: B. It varies the frequency of the supply to the motor, allowing variable speed control.

Explanation: A frequency inverter benefits a motor control system by varying the frequency of the electrical supply to the motor. This allows for adjustable speed control, enhancing the

155

system's adaptability and efficiency in various operating conditions.

Answer 358: C. It protects the motor from overheating due to excessive current.

Explanation: A thermal overload relay offers protection to a motor by sensing excessive current that may cause overheating. It automatically cuts off power to the motor when overheating is detected, preventing damage and ensuring operational safety.

Answer 359: A. To reduce the mechanical stress on motor and driven equipment during startup.

Explanation: A motor soft starter gradually ramps up the power supply to the motor, reducing mechanical stress and electrical inrush current during startup, which enhances the longevity and reliability of both the motor and the connected equipment.

Answer 360: B. Reversing contactor.

Explanation: A reversing contactor is essential in automated systems for reversing the direction of a DC motor. It switches the connections to the motor's windings to change the rotation direction efficiently.

Answer 361: C. By automatically switching the power source to maintain continuous operation.

Explanation: An Automatic Transfer Switch (ATS) is used in motor control systems to automatically switch between power sources without manual intervention, ensuring continuous motor operation during power outages or source fluctuations.

Answer 362: C. To detect the physical movement or position of an object.

Explanation: Limit switches in automation systems are used to detect the presence, position, or movement of an object. They act as safety devices that halt machinery when movement exceeds preset boundaries, ensuring safe operation.

Answer 363: C. Via electronic control, such as with a VFD that adjusts the power supplied to the motor.

Explanation: Torque control in motor systems is primarily achieved through electronic means such as Variable Frequency Drives (VFDs), which adjust the power input to control the torque output of the motor, optimizing performance and protecting against mechanical stress.

Answer 364: B. Built-in circuit breakers.

156

Explanation: Modern motor starters often incorporate built-in circuit breakers to provide immediate protection against electrical faults, safeguarding the motor and circuit from damage due to overcurrent or short circuits.

Answer 365: C. They control the sequence of operations and automate processes.

Explanation: Programmable Logic Controllers (PLCs) are critical in automated motor control systems for controlling the sequence of operations. They automate complex processes, ensuring high precision, repeatability, and efficiency in industrial settings.

Answer 366: B. It provides precise control over the motor's position, speed, and acceleration.

Explanation: Encoders are used with servo motors to provide real-time feedback on position, speed, and acceleration, enabling precise control necessary for applications requiring exact motion and positioning.

Answer 367: C. To ensure safety by dissipating fault currents.

Explanation: Grounding in motor control panels is crucial for safety as it helps dissipate fault currents safely into the earth, preventing potential electric shocks and ensuring safe operation and maintenance.

Answer 368: B. To display process data and facilitate user control over the system.

Explanation: Human-Machine Interfaces (HMIs) in industrial automation systems are essential for displaying process data and allowing operators to control and monitor the system efficiently, enhancing interaction between human operators and machine processes.

Answer 369: B. By detecting excessive heat and interrupting the power supply.

Explanation: Thermal protectors in motors are designed to detect excessive heat and automatically interrupt the power supply to the motor, preventing potential damage due to overheating.

Answer 370: A. It uses a magnetic field to mechanically switch the motor circuit on or off.

Explanation: Magnetic contactors use a magnetic field to actuate a set of mechanical contacts that switch the motor circuit on or off. This allows for remote control of large motors and reduces the physical wear associated with mechanical switching.

Answer 371: B. Temperature and vibration sensors.

Explanation: In motor control systems, temperature and vibration sensors are commonly used to monitor the condition of motors, ensuring efficient operation and preventing damage by detecting and addressing potential issues early.

Answer 372: B. To convert DC to AC, enabling precise speed control of AC motors.

Explanation: Inverters are used in motor control applications to convert DC power from the power supply to AC, which can be precisely controlled to adjust the speed of AC motors, making them versatile for various applications.

Answer 373: B. By detecting phase loss, phase imbalance, and phase reversal to prevent motor damage.

Explanation: Phase monitors in industrial environments protect motors by detecting issues like phase loss, phase imbalance, and phase reversal. This protection is critical to prevent damage to motors that can occur if these conditions are left unchecked.

Answer 374: B. To centralize control of various motors, providing power, control, and monitoring capabilities.

Explanation: A Motor Control Center (MCC) is crucial in large industrial plants for centralizing the control of various motors. It provides a consolidated point for power distribution, control, and monitoring, enhancing safety and efficiency.

Answer 375: D. Large, three-phase motors.

Explanation: Direct online starters are typically used with large, three-phase motors because they provide a simple and effective method to start these motors by directly connecting them to the power supply, suitable for motors that do not require speed control or gradual starting.

Answer 376: C. To improve power factor and provide necessary phase shifts for starting torque.

Explanation: Capacitors are used with electric motors to improve the power factor of the motor circuit and provide necessary phase shifts required for creating starting torque, enhancing the efficiency and effectiveness of motor operations.

Answer 377: B. Power, speed, torque, and the control system compatibility.

Explanation: When selecting a motor for an automated system, considerations such as the motor's power capacity, speed, torque, and compatibility with the existing control system are critical to ensure optimal performance and integration.

Answer 378: B. By changing the frequency of the input pulses.

Explanation: Speed control in stepper motors is achieved by varying the frequency of the input pulses supplied to the motor. This allows for precise control of the motor speed, essential for applications requiring exact positioning.

Answer 379: B. They protect the motor from excessive current by breaking the circuit when necessary.

Explanation: Overload relays are used in motor control circuits to protect the motor from excessive current. They detect overcurrent conditions and break the circuit to prevent motor damage and potential fire hazards.

Answer 380: B. By interrupting power flow in case of an overload, preventing potential motor damage or electrical fires.

Explanation: Circuit breakers in motor control panels enhance safety by automatically interrupting the power flow in the event of an overload or short circuit, preventing motor damage and electrical fires, and ensuring a safe operating environment.

Answer 381: A. It allows for speed and position control that is in sync with the power grid frequency.

Explanation: A synchronous motor controller is designed to allow the motor to operate in sync with the power grid frequency, enabling precise speed and position control, which is critical for applications requiring exact synchronization.

Answer 382: A. To reduce the need for human intervention, increase efficiency, and improve system reliability.

Explanation: Integrating automation systems with motor controls reduces the need for manual intervention, enhances operational efficiency, and improves the overall reliability of the system, leading to increased productivity and safety in industrial environments.

Electrical Feeders

Q383: What is the primary function of electrical feeders?

A. To regulate voltage in a circuit
B. To distribute power from one main source to various branch circuits
C. To convert AC to DC
D. To store electrical energy

Q384: What factor must be considered when sizing an electrical feeder for a new installation?

A. Color of the feeder cable
B. Total expected load of the circuit
C. Length of the cable only
D. Type of insulation material

Q385: How does the presence of harmonics in the system affect feeder cable selection?

A. No effect, as harmonics are unrelated to feeders
B. Requires thicker insulation to prevent interference
C. Might need cables with higher current carrying capacity to handle increased heat
D. Shorter cable lengths are preferred to minimize effects

Q386: What is the significance of the impedance in feeder cables?

A. Lower impedance helps in reducing energy costs
B. High impedance is necessary for better voltage regulation
C. Impedance impacts the voltage drop along the cable
D. Impedance affects the color and texture of the cable

Q387: Why is it crucial to ensure proper grounding in electrical feeder systems?

A. Grounding affects the operational speed of the circuit
B. Proper grounding prevents electrical shock and stabilizes voltage levels
C. Grounding is only necessary for aesthetic purposes
D. Grounding increases the energy efficiency of the system

Q388: What is a feeder pillar and what role does it play in electrical distribution systems?

A. A decorative component of the electrical system
B. A central point for connecting different feeder cables
C. An advanced sensor for detecting current flow
D. A device that steps up the voltage in feeder cables

Q389: How are feeder lines protected from overcurrent and short circuits?

A. By using fuses and circuit breakers rated for the expected loads
B. By installing additional capacitors along the line

C. By decreasing the wire diameter
D. By manually monitoring the feeders continuously

Q390: What is voltage drop in feeder cables and why is it important to control it?

A. It enhances the efficiency of the electrical system
B. Voltage drop refers to the reduction in voltage as electricity travels along the cable, affecting how devices function at the circuit's end
C. Voltage drop is a desired characteristic in residential installations
D. It refers to the physical sagging of cables over time

Q391: What are the consequences of improperly sized feeder cables?

A. Increased aesthetic appeal of the electrical system
B. Can lead to excessive voltage drop, overheating, and potential fire hazards
C. Makes the system more energy-efficient
D. Reduces the cost of materials

Q392: How should feeder cables be installed to ensure safety and compliance with electrical codes?

A. Loosely to allow for natural expansion and contraction
B. In a straight line regardless of the landscape
C. According to manufacturer's instructions and local code requirements, often in conduits or cable trays
D. With minimal securing to allow for easy re-routing

Q393: What is the role of a feeder breaker in an electrical distribution system?

A. To break the cable for easy installation
B. To manually turn circuits on and off
C. To interrupt power automatically in case of an overload or short-circuit
D. To provide a fixed connection point between different cable types

Q394: In what scenarios might a feeder need to be derated?

A. In low-temperature environments
B. When the feeder is exposed to high ambient temperatures or is bundled

161

with other cables, potentially increasing heat buildup
C. When the feeder is under low load conditions
D. When the feeder is used in temporary installations

Q395: What considerations should be made for the expansion and contraction of feeder cables?

A. Cables should be painted to prevent expansion
B. Allowance for thermal expansion should be made in cable runs to prevent mechanical stress
C. Cables should be tightly clamped to prevent movement
D. Expansion is not a concern with modern cables

Q396: How does the choice of conductor material impact the performance of feeder cables?

A. Only gold conductors should be used for optimal performance
B. Copper offers lower resistance than aluminum, which can affect efficiency and voltage drop
C. Conductor material has no impact on performance
D. All conductors perform the same under electrical load

Q397: What safety tests should be conducted on feeder cables before they are put into service?

A. Color testing to ensure they are visible
B. Insulation resistance tests and continuity checks to ensure there are no defects
C. Weight tests to ensure they can support physical loads
D. Aesthetic checks to ensure they match the environment

Q398: How does the length of a feeder cable affect its design and specification?

A. Longer cables require lighter colors to reduce heat absorption
B. Longer feeder cables might require larger cross-sectional areas to mitigate voltage drop and power loss
C. Length has no impact on the design of feeder cables
D. Shorter cables always require more insulation

Q399: What is the impact of feeder cable capacitance in high-frequency power transmission?

A. It increases the speed of electricity flow
B. Capacitance can lead to power losses and affect voltage regulation
C. Capacitance is only a concern in low-frequency applications
D. It makes the system more cost-effective

Q400: Why must feeder cables be resistant to UV light and chemicals?

A. UV and chemical resistance is only necessary for indoor cables
B. It ensures that the cables do not conduct electricity
C. Resistance to these elements prevents degradation and prolongs the life of the cable in harsh environments
D. It affects the color and design of the cables

Q401: How do environmental considerations affect the routing of feeder cables?

A. Cables must be routed to be aesthetically pleasing
B. Environmental factors like potential flooding, chemical exposure, or mechanical damage risk must be considered to ensure reliability
C. Routing is typically planned to create the shortest possible path, regardless of environment
D. Environmental considerations are secondary to cost considerations

Q402: What type of insulation is typically used for feeder cables in industrial applications?

A. Basic plastic insulation
B. High-grade insulation materials like cross-linked polyethylene (XLPE) for thermal and electrical resistance
C. Decorative insulation
D. Insulation is not required for industrial feeder cables

Q403: What are the typical maintenance routines for electrical feeders in a commercial building?

A. Daily cleaning and polishing
B. Regular inspections for wear and damage, and periodic testing for electrical performance
C. Maintenance routines are not necessary for modern systems
D. Only replaced when visible faults occur

Q404: How does the installation of feeder cables in public areas differ from private areas?

A. Public installations require more decorative cables
B. Public installations must adhere to stricter safety standards and are often required to be buried or placed in protective conduits
C. There is no difference in the installation process
D. Private installations require higher security measures

Q405: What is a feeder tap, and what are the regulations that must be considered when implementing one?

A. A decorative element in feeder systems
B. A connection to the main feeder cable that supplies power to a smaller branch circuit, subject to specific code requirements regarding its size and length
C. A type of switch used to control feeder lines
D. A monitoring device installed on feeders

Q406: How should feeder cables be rated in terms of current capacity relative to the expected load?

A. At exactly the same current as the expected load for efficiency
B. Lower than the expected load to save on costs
C. Higher than the expected load to allow for future expansion and avoid overheating
D. Randomly, as it does not affect performance

Q407: What role does the conductor's temperature rating play in the selection of feeder cables?

A. The temperature rating indicates the aesthetic value of the cable
B. It is crucial for determining the safe operating temperature range of the cable
C. Temperature rating is only important for outdoor cables
D. It determines the electrical resistance of the cable

Q408: In what scenarios are underground feeder (UF) cables most appropriately used?

A. In situations requiring minimal visual impact and protection from environmental factors
B. Underground cables are only used for temporary setups

164

C. UF cables are typically used indoors
D. They are most commonly used in overhead installations

Q409: How does the presence of feeder cables impact the design of electrical panels?

A. It does not impact panel design
B. Feeders require panels to have higher aesthetic qualities
C. Panels must accommodate the physical and electrical characteristics of feeder cables, including space for safe routing and connections
D. Panels are smaller when feeders are used

Q410: What testing procedures should be followed for feeder cables after installation to ensure they meet safety standards?

A. No testing is necessary
B. Only visual inspections are required
C. Electrical testing such as insulation resistance and continuity tests
D. Only color testing to ensure proper coding

Q411: How should feeder cables be protected from mechanical damage, especially in industrial settings?

A. By covering them with decorative tape
B. By using robust conduit systems or armored cable where risk of damage is high
C. Protection is not necessary if cables are installed professionally
D. By painting them with protective coatings

Q412: What is the impact of long feeder runs on power quality, and how can it be mitigated?

A. Longer runs improve power quality by increasing voltage
B. Long runs may lead to voltage drops and increased impedance, mitigated by using cables with larger cross-sectional areas or installing repeaters
C. Long feeder runs have no impact on power quality
D. They decrease power quality by reducing current

Q413: How does the orientation of feeder cables affect their performance in electromagnetic interference (EMI) heavy environments?

A. Orientation is irrelevant to EMI effects
B. Specific orientation and shielding strategies can help minimize EMI effects
C. Cables should be oriented vertically only
D. Horizontal orientation minimizes EMI

Q414: What considerations must be made for feeder cables in areas prone to seismic activity?

A. No special considerations are needed
B. Use of flexible supports and expansion joints to allow for movement without damage
C. Cables should be loosely laid to provide flexibility
D. Heavier insulation should be used

Q415: How do environmental factors like humidity and salinity affect the choice of feeder cable materials?

A. They necessitate the use of gold-plated cables
B. Resistant materials such as tinned copper or moisture-resistant insulation are chosen to prevent corrosion and degradation
C. These factors enhance the conductivity of cables
D. Humidity and salinity are primarily concerns for wireless transmission

Q416: What advancements in feeder cable technology are improving safety and efficiency in modern electrical grids?

A. The use of fiber optics only
B. Innovations like smart grid-compatible cables, higher temperature insulation materials, and improved conductor materials
C. The shift from AC to DC systems
D. The use of lighter colors to reduce heat absorption

Q417: What is the recommended practice for labeling feeder cables in a large facility?

A. Labels are not necessary if cables are color-coded
B. Use generic labels for ease of installation
C. Detailed labeling including source, destination, and power specifications to facilitate maintenance and troubleshooting
D. Using symbols only for quick identification

Q418: How can the reliability of feeder cable connections be ensured over time?

A. By reducing the number of connections
B. Regular maintenance checks, proper installation practices, and the use of suitable connectors and joints
C. Reliability is naturally ensured by using high-quality cables
D. By tightening connections annually

Q419: What are the challenges of integrating renewable energy sources with existing feeder cable infrastructure?

A. No challenges exist; integration is seamless
B. Managing variable power outputs and ensuring compatibility with grid standards
C. Renewable sources are less powerful and do not affect feeders
D. The main challenge is aesthetic integration

Q420: How should feeder cables be routed to minimize interference with communication lines?

A. Parallel and close to communication lines to save space
B. At a safe distance or using shielded cables to prevent electromagnetic interference
C. Interference is not an issue with modern cables
D. Routing them randomly, as it does not matter

Q421: What are the specific code requirements that dictate the use and installation of feeder cables in residential versus commercial buildings?

A. Residential installations require less stringent codes
B. Both types of buildings follow the same codes for simplicity
C. Codes vary significantly, typically requiring more robust protection and redundancy in commercial settings
D. Commercial buildings do not require adherence to safety codes

Q422: How do feeder cables influence the overall energy efficiency of an electrical system?

A. They have no influence on energy efficiency
B. Properly sized and installed feeder cables reduce losses and improve system efficiency
C. Feeder cables decrease efficiency as they age
D. Only new cables improve efficiency

Q423: What are the consequences of failing to periodically inspect and maintain electrical feeder systems?

A. Systems become more efficient over time
B. Potential for system failures, safety hazards, and inefficient operation
C. Periodic inspections can damage the cables
D. Maintenance only affects the visual appeal of the system

Q424: What technologies are being implemented to monitor the condition of feeder cables in real time?

A. Basic visual inspection technologies
B. Advanced sensor networks that detect and report changes in electrical characteristics
C. No technologies are available for real-time monitoring
D. Use of manual checking methods only

Q425: How do changes in regulatory standards affect the installation and upgrade of feeder systems?

A. Standards have minimal impact on existing installations
B. New standards often require significant updates to systems to meet safety, efficiency, and environmental benchmarks
C. Regulatory changes only affect new installations
D. Changes in standards are only advisory, not mandatory

Q426: What are the best practices for decommissioning old feeder cables?

A. Leave them in place to maintain historical integrity
B. Safe removal, recycling of materials, and ensuring that new installations do not disturb the decommissioned cables
C. Reuse in different applications within the same facility
D. Burning them to ensure complete disposal

Q427: What type of protective device is typically used to prevent overloading in electrical feeder systems?

A. Grounding rod
B. Circuit breaker
C. Voltage regulator
D. Frequency modulator

Q428: In what situations is the use of aluminum conductors in feeder cables preferable to copper?

A. When higher conductivity is required
B. When lower cost and lighter weight are beneficial
C. When maximum flexibility is needed
D. When higher resistance to corrosion is necessary

Answers

Answer 383: B. To distribute power from one main source to various branch circuits.

Explanation: The primary function of electrical feeders is to carry and distribute electrical power from a main power source to various branch circuits, ensuring efficient power distribution within a building or system.

Answer 384: B. Total expected load of the circuit.

Explanation: When sizing an electrical feeder for a new installation, the total expected load of the circuit must be considered to ensure the feeder can handle the power demand without overheating or causing voltage drops.

Answer 385: C. Might need cables with higher current carrying capacity to handle increased heat.

Explanation: The presence of harmonics in the system can increase the heat generated in feeder cables, which may require selecting cables with higher current carrying capacity to ensure safe and reliable operation.

Answer 386: C. Impedance impacts the voltage drop along the cable.

Explanation: The impedance of feeder cables affects the voltage drop along the cable. Lower impedance reduces voltage drop, ensuring that devices at the end of the feeder receive adequate voltage for proper operation.

Answer 387: B. Proper grounding prevents electrical shock and stabilizes voltage levels.

Explanation: Ensuring proper grounding in electrical feeder systems is crucial for preventing electrical shock hazards and stabilizing voltage levels, which enhances the safety and reliability of the electrical system.

Answer 388: B. A central point for connecting different feeder cables.

Explanation: A feeder pillar is a crucial component in electrical distribution systems, serving as a central point for connecting and distributing power through different feeder cables.

Answer 389: A. By using fuses and circuit breakers rated for the expected loads.

Explanation: Feeder lines are protected from overcurrent and short circuits by using fuses and circuit breakers that are appropriately rated for the expected electrical loads, ensuring safe and reliable operation.

Answer 390: B. Voltage drop refers to the reduction in voltage as electricity travels along the cable, affecting how devices function at the circuit's end.

Explanation: Voltage drop in feeder cables is the reduction in voltage as electricity travels along the cable, which can impact the performance of devices at the circuit's end. Controlling voltage drop is important to ensure devices operate correctly.

Answer 391: B. Can lead to excessive voltage drop, overheating, and potential fire hazards.

Explanation: Improperly sized feeder cables can lead to excessive voltage drop, overheating, and potential fire hazards, compromising the safety and efficiency of the electrical system.

Answer 392: C. According to manufacturer's instructions and local code requirements, often in conduits or cable trays.

Explanation: Feeder cables should be installed according to the manufacturer's instructions and local code requirements, often in conduits or cable trays, to ensure safety and compliance with electrical codes.

Answer 393: C. To interrupt power automatically in case of an overload or short-circuit.

Explanation: A feeder breaker is designed to automatically interrupt power in the event of an overload or short circuit, protecting the electrical distribution system from damage and enhancing safety.

Answer 394: B. When the feeder is exposed to high ambient temperatures or is bundled with other cables, potentially increasing heat buildup.

Explanation: A feeder might need to be derated when exposed to high ambient temperatures or bundled with other cables, as these conditions can increase heat buildup and affect the feeder's current carrying capacity.

Answer 395: B. Allowance for thermal expansion should be made in cable runs to prevent mechanical stress.

Explanation: When installing feeder cables, allowances for thermal expansion should be made to prevent mechanical stress and ensure the cables can expand and contract without causing damage.

Answer 396: B. Copper offers lower resistance than aluminum, which can affect efficiency and voltage drop.

Explanation: The choice of conductor material, such as copper or aluminum, impacts the performance of feeder cables. Copper offers lower resistance than aluminum, which can improve efficiency and reduce voltage drop.

Answer 397: B. Insulation resistance tests and continuity checks to ensure there are no defects.

Explanation: Before putting feeder cables into service, insulation resistance tests and continuity checks should be conducted to ensure there are no defects and the cables are safe for operation.

Answer 398: B. Longer feeder cables might require larger cross-sectional areas to mitigate voltage drop and power loss.

Explanation: The length of a feeder cable affects its design and specification, as longer cables might require larger cross-sectional areas to mitigate voltage drop and power loss, ensuring efficient power transmission.

Answer 399: B. Capacitance can lead to power losses and affect voltage regulation.

Explanation: In high-frequency power transmission, the capacitance of feeder cables can lead to power losses and affect voltage regulation, which needs to be managed to maintain system efficiency.

Answer 400: C. Resistance to these elements prevents degradation and prolongs the life of the cable in harsh environments.

Explanation: Feeder cables must be resistant to UV light and chemicals to prevent degradation and prolong their life in harsh environments, ensuring reliable and safe operation over time.

Answer 401: B. Environmental factors like potential flooding, chemical exposure, or mechanical damage risk must be considered to ensure reliability.

Explanation: Routing feeder cables must consider environmental factors such as potential flooding, chemical exposure, or mechanical damage to ensure the reliability and longevity of the electrical system.

Answer 402: B. High-grade insulation materials like cross-linked polyethylene (XLPE) for thermal and electrical resistance.

Explanation: In industrial applications, feeder cables typically use high-grade insulation materials like cross-linked polyethylene (XLPE) for their superior thermal and electrical resistance properties.

Answer 403: B. Regular inspections for wear and damage, and periodic testing for electrical performance.

Explanation: Maintenance routines for electrical feeders in commercial buildings include regular inspections for wear and damage, and periodic testing for electrical performance to ensure safety and efficiency.

Answer 404: B. Public installations must adhere to stricter safety standards and are often required to be buried or placed in protective conduits.

Explanation: The installation of feeder cables in public areas must adhere to stricter safety standards and often requires cables to be buried or placed in protective conduits to prevent accidents and ensure public safety.

Answer 405: B. A connection to the main feeder cable that supplies power to a smaller branch circuit, subject to specific code requirements regarding its size and length.

Explanation: A feeder tap is a connection to the main feeder cable that supplies power to a smaller branch circuit and must comply with specific code requirements regarding its size and length to ensure safe operation.

Answer 406: C. Higher than the expected load to allow for future expansion and avoid overheating.

Explanation: Feeder cables should be rated higher than the expected load to allow for future expansion and to avoid overheating, ensuring the safe and efficient operation of the electrical system.

Answer 407: B. It is crucial for determining the safe operating temperature range of the cable.

Explanation: The temperature rating of a conductor is crucial for determining the safe operating temperature range of the cable, ensuring it can handle the expected electrical load without degrading.

Answer 408: A. In situations requiring minimal visual impact and protection from environmental factors.

Explanation: Underground feeder (UF) cables are most appropriately used in situations requiring minimal visual impact and protection from environmental factors, providing a safe and unobtrusive solution.

Answer 409: C. Panels must accommodate the physical and electrical characteristics of feeder cables, including space for safe routing and connections.

Explanation: Electrical panels must be designed to accommodate the physical and electrical characteristics of feeder cables, ensuring there is adequate space for safe routing and secure connections.

Answer 410: C. Electrical testing such as insulation resistance and continuity tests.

Explanation: After installation, feeder cables should undergo electrical testing such as insulation resistance and continuity tests to ensure they meet safety standards and are free from defects.

Answer 411: B. By using robust conduit systems or armored cable where risk of damage is high.

Explanation: Feeder cables in industrial settings should be protected from mechanical damage by using robust conduit systems or armored cables, especially in areas where the risk of damage is high.

Answer 412: B. Long runs may lead to voltage drops and increased impedance, mitigated by using cables with larger cross-sectional areas or installing repeaters.

Explanation: Long feeder runs can lead to voltage drops and increased impedance, which can be mitigated by using cables with larger cross-sectional areas or by installing repeaters to boost the signal.

Answer 413: B. Specific orientation and shielding strategies can help minimize EMI effects.

Explanation: The orientation of feeder cables and the use of shielding strategies can help minimize the effects of electromagnetic interference (EMI) in environments with high EMI levels.

Answer 414: B. Use of flexible supports and expansion joints to allow for movement without damage.

Explanation: In areas prone to seismic activity, feeder cables should be installed with flexible supports and expansion joints to allow for movement without causing damage to the cables.

Answer 415: B. Resistant materials such as tinned copper or moisture-resistant insulation are chosen to prevent corrosion and degradation.

Explanation: In environments with high humidity and salinity, materials like tinned copper and moisture-resistant insulation are chosen for feeder cables to prevent corrosion and degradation.

Answer 416: B. Innovations like smart grid-compatible cables, higher temperature insulation materials, and improved conductor materials.

Explanation: Advancements in feeder cable technology, such as smart grid-compatible cables, higher temperature insulation materials, and improved conductor materials, are improving safety and efficiency in modern electrical grids.

Answer 417: C. Detailed labeling including source, destination, and power specifications to facilitate maintenance and troubleshooting.

Explanation: Detailed labeling of feeder cables, including source, destination, and power specifications, is recommended to facilitate maintenance and troubleshooting in large facilities.

Answer 418: B. Regular maintenance checks, proper installation practices, and the use of suitable connectors and joints.

Explanation: The reliability of feeder cable connections can be ensured through regular maintenance checks, proper installation practices, and the use of suitable connectors and joints.

Answer 419: B. Managing variable power outputs and ensuring compatibility with grid standards.

Explanation: Integrating renewable energy sources with existing feeder cable infrastructure involves managing variable power outputs and ensuring compatibility with grid standards.

Answer 420: B. At a safe distance or using shielded cables to prevent electromagnetic interference.

Explanation: Feeder cables should be routed at a safe distance from communication lines or use shielded cables to prevent electromagnetic interference.

Answer 421: C. Codes vary significantly, typically requiring more robust protection and redundancy in commercial settings.

Explanation: Code requirements for feeder cables differ between residential and commercial buildings, with commercial settings typically requiring more robust protection and redundancy.

Answer 422: B. Properly sized and installed feeder cables reduce losses and improve system efficiency.

Explanation: Properly sized and installed feeder cables help reduce electrical losses and improve the overall energy efficiency of the electrical system.

Answer 423: B. Potential for system failures, safety hazards, and inefficient operation.

Explanation: Failing to periodically inspect and maintain electrical feeder systems can lead to system failures, safety hazards, and inefficient operation.

Answer 424: B. Advanced sensor networks that detect and report changes in electrical characteristics.

Explanation: Technologies like advanced sensor networks are being implemented to monitor the condition of feeder cables in real-time, detecting and reporting changes in electrical characteristics.

Answer 425: B. New standards often require significant updates to systems to meet safety, efficiency, and environmental benchmarks.

Explanation: Changes in regulatory standards often require significant updates to electrical systems to meet new safety, efficiency, and environmental benchmarks.

Answer 426: B. Safe removal, recycling of materials, and ensuring that new installations do not disturb the decommissioned cables.

Explanation: The best practices for decommissioning old feeder cables include safe removal, recycling of materials, and ensuring that new installations do not disturb the decommissioned cables.

Answer 427: B. Circuit breaker.

Explanation: A circuit breaker is typically used as a protective device to prevent overloading in electrical feeder systems.

Answer 428: B. When lower cost and lighter weight are beneficial.

Explanation: The use of aluminum conductors in feeder cables is preferable to copper when lower cost and lighter weight are beneficial, although aluminum has higher resistance than copper.

Complete Exam

Q1: What is the primary function of electrical feeders?

    A. To regulate voltage in a circuit
    B. To distribute power from one main source to various branch circuits
    C. To convert AC to DC
    D. To store electrical energy

Q2: What factor must be considered when sizing an electrical feeder for a new installation?

A. Color of the feeder cable
B. Total expected load of the circuit
C. Length of the cable only
D. Type of insulation material

Q3: How does the presence of harmonics in the system affect feeder cable selection?

A. No effect, as harmonics are unrelated to feeders
B. Requires thicker insulation to prevent interference
C. Might need cables with higher current carrying capacity to handle increased heat
D. Shorter cable lengths are preferred to minimize effects

Q4: What is the significance of the impedance in feeder cables?

A. Lower impedance helps in reducing energy costs
B. High impedance is necessary for better voltage regulation
C. Impedance impacts the voltage drop along the cable
D. Impedance affects the color and texture of the cable

Q5: Why is it crucial to ensure proper grounding in electrical feeder systems?

A. Grounding affects the operational speed of the circuit
B. Proper grounding prevents electrical shock and stabilizes voltage levels
C. Grounding is only necessary for aesthetic purposes
D. Grounding increases the energy efficiency of the system

Q6: What is a feeder pillar and what role does it play in electrical distribution systems?

A. A decorative component of the electrical system
B. A central point for connecting different feeder cables
C. An advanced sensor for detecting current flow
D. A device that steps up the voltage in feeder cables

Q7: How are feeder lines protected from overcurrent and short circuits?

A. By using fuses and circuit breakers rated for the expected loads

B. By installing additional capacitors along the line
C. By decreasing the wire diameter
D. By manually monitoring the feeders continuously

Q8: What is voltage drop in feeder cables and why is it important to control it?

A. It enhances the efficiency of the electrical system
B. Voltage drop refers to the reduction in voltage as electricity travels along the cable, affecting how devices function at the circuit's end
C. Voltage drop is a desired characteristic in residential installations
D. It refers to the physical sagging of cables over time

Q9: What are the consequences of improperly sized feeder cables?

A. Increased aesthetic appeal of the electrical system
B. Can lead to excessive voltage drop, overheating, and potential fire hazards
C. Makes the system more energy-efficient
D. Reduces the cost of materials

Q10: How should feeder cables be installed to ensure safety and compliance with electrical codes?

A. Loosely to allow for natural expansion and contraction
B. In a straight line regardless of the landscape
C. According to manufacturer's instructions and local code requirements, often in conduits or cable trays
D. With minimal securing to allow for easy re-routing

Q11: What is the role of a feeder breaker in an electrical distribution system?

A. To break the cable for easy installation
B. To manually turn circuits on and off
C. To interrupt power automatically in case of an overload or short-circuit
D. To provide a fixed connection point between different cable types

Q12: In what scenarios might a feeder need to be derated?

In low-temperature environments

A. When the feeder is exposed to high ambient temperatures or is bundled

177

with other cables, potentially increasing heat buildup
B. When the feeder is under low load conditions
C. When the feeder is used in temporary installations

Q13: What considerations should be made for the expansion and contraction of feeder cables?

A. Cables should be painted to prevent expansion
B. Allowance for thermal expansion should be made in cable runs to prevent mechanical stress
C. Cables should be tightly clamped to prevent movement
D. Expansion is not a concern with modern cables

Q14: How does the choice of conductor material impact the performance of feeder cables?

A. Only gold conductors should be used for optimal performance
B. Copper offers lower resistance than aluminum, which can affect efficiency and voltage drop
C. Conductor material has no impact on performance
D. All conductors perform the same under electrical load

Q15: What safety tests should be conducted on feeder cables before they are put into service?

A. Color testing to ensure they are visible
B. Insulation resistance tests and continuity checks to ensure there are no defects
C. Weight tests to ensure they can support physical loads
D. Aesthetic checks to ensure they match the environment

Q16: How does the length of a feeder cable affect its design and specification?

A. Longer cables require lighter colors to reduce heat absorption
B. Longer feeder cables might require larger cross-sectional areas to mitigate voltage drop and power loss
C. Length has no impact on the design of feeder cables
D. Shorter cables always require more insulation

Q17: What is the impact of feeder cable capacitance in high-frequency power transmission?

A. It increases the speed of electricity flow
B. Capacitance can lead to power losses and affect voltage regulation
C. Capacitance is only a concern in low-frequency applications
D. It makes the system more cost-effective

Q18: Why must feeder cables be resistant to UV light and chemicals?

A. UV and chemical resistance is only necessary for indoor cables
B. It ensures that the cables do not conduct electricity
C. Resistance to these elements prevents degradation and prolongs the life of the cable in harsh environments
D. It affects the color and design of the cables

Q19: How do environmental considerations affect the routing of feeder cables?

A. Cables must be routed to be aesthetically pleasing
B. Environmental factors like potential flooding, chemical exposure, or mechanical damage risk must be considered to ensure reliability
C. Routing is typically planned to create the shortest possible path, regardless of environment
D. Environmental considerations are secondary to cost considerations

Q20: What type of insulation is typically used for feeder cables in industrial applications?

A. Basic plastic insulation
B. High-grade insulation materials like cross-linked polyethylene (XLPE) for thermal and electrical resistance
C. Decorative insulation
D. Insulation is not required for industrial feeder cables

Q21: What are the typical maintenance routines for electrical feeders in a commercial building?

A. Daily cleaning and polishing
B. Regular inspections for wear and damage, and periodic testing for electrical performance
C. Maintenance routines are not necessary for modern systems
D. Only replaced when visible faults occur

Q22: How does the installation of feeder cables in public areas differ from private areas?

A. Public installations require more decorative cables
B. Public installations must adhere to stricter safety standards and are often required to be buried or placed in protective conduits
C. There is no difference in the installation process
D. Private installations require higher security measures

Q23: What is a feeder tap, and what are the regulations that must be considered when implementing one?

A. A decorative element in feeder systems
B. A connection to the main feeder cable that supplies power to a smaller branch circuit, subject to specific code requirements regarding its size and length
C. A type of switch used to control feeder lines
D. A monitoring device installed on feeders

Q24: How should feeder cables be rated in terms of current capacity relative to the expected load?

A. At exactly the same current as the expected load for efficiency
B. Lower than the expected load to save on costs
C. Higher than the expected load to allow for future expansion and avoid overheating
D. Randomly, as current capacity is not critical

Q25: Why is the temperature rating of a conductor important in the context of feeder cables?

A. It determines the physical flexibility of the cable
B. It is crucial for determining the safe operating temperature range of the cable
C. It affects the color of the insulation material
D. It is only relevant for high-voltage applications

Q26: In what scenarios are underground feeder (UF) cables most appropriately used?

A. In situations requiring minimal visual impact and protection from environmental factors
B. Underground cables are only used for temporary setups

C. UF cables are typically used indoors
D. They are most commonly used in overhead installations

Q27: How does the presence of feeder cables impact the design of electrical panels?

A. It does not impact panel design
B. Feeders require panels to have higher aesthetic qualities
C. Panels must accommodate the physical and electrical characteristics of feeder cables, including space for safe routing and connections
D. Panels are smaller when feeders are used

Q28: What testing procedures should be followed for feeder cables after installation to ensure they meet safety standards?

A. No testing is necessary
B. Only visual inspections are required
C. Electrical testing such as insulation resistance and continuity tests
D. Only color testing to ensure proper coding

Q29: How should feeder cables be protected from mechanical damage, especially in industrial settings?

A. By covering them with decorative tape
B. By using robust conduit systems or armored cable where risk of damage is high
C. Protection is not necessary if cables are installed professionally
D. By painting them with protective coatings

Q30: What is the impact of long feeder runs on power quality, and how can it be mitigated?

A. Longer runs improve power quality by increasing voltage
B. Long runs may lead to voltage drops and increased impedance, mitigated by using cables with larger cross-sectional areas or installing repeaters
C. Long feeder runs have no impact on power quality
D. They decrease power quality by reducing current

Q31: How does the orientation of feeder cables affect their performance in electromagnetic interference (EMI) heavy environments?

181

A. Orientation is irrelevant to EMI effects
B. Specific orientation and shielding strategies can help minimize EMI effects
C. Cables should be oriented vertically only
D. Horizontal orientation minimizes EMI

Q32: What considerations must be made for feeder cables in areas prone to seismic activity?

A. No special considerations are needed
B. Use of flexible supports and expansion joints to allow for movement without damage
C. Cables should be loosely laid to provide flexibility
D. Heavier insulation should be used

Q33: How do environmental factors like humidity and salinity affect the choice of feeder cable materials?

A. They necessitate the use of gold-plated cables
B. Resistant materials such as tinned copper or moisture-resistant insulation are chosen to prevent corrosion and degradation
C. These factors enhance the conductivity of cables
D. Humidity and salinity are primarily concerns for wireless transmission

Q34: What advancements in feeder cable technology are improving safety and efficiency in modern electrical grids?

A. The use of fiber optics only
B. Innovations like smart grid-compatible cables, higher temperature insulation materials, and improved conductor materials
C. The shift from AC to DC systems
D. The use of lighter colors to reduce heat absorption

Q35: What is the recommended practice for labeling feeder cables in a large facility?

A. Labels are not necessary if cables are color-coded
B. Use generic labels for ease of installation
C. Detailed labeling including source, destination, and power specifications to facilitate maintenance and troubleshooting
D. Using symbols only for quick identification

Q36: How can the reliability of feeder cable connections be ensured over time?

A. By reducing the number of connections
B. Regular maintenance checks, proper installation practices, and the use of suitable connectors and joints
C. Reliability is naturally ensured by using high-quality cables
D. By tightening connections annually

Q37: What are the challenges of integrating renewable energy sources with existing feeder cable infrastructure?

A. No challenges exist; integration is seamless
B. Managing variable power outputs and ensuring compatibility with grid standards
C. Renewable sources are less powerful and do not affect feeders
D. The main challenge is aesthetic integration

Q38: How should feeder cables be routed to minimize interference with communication lines?

A. Parallel and close to communication lines to save space
B. At a safe distance or using shielded cables to prevent electromagnetic interference
C. Interference is not an issue with modern cables
D. Routing them randomly, as it does not matter

Q39: What are the specific code requirements that dictate the use and installation of feeder cables in residential versus commercial buildings?

A. Residential installations require less stringent codes
B. Both types of buildings follow the same codes for simplicity
C. Codes vary significantly, typically requiring more robust protection and redundancy in commercial settings
D. Commercial buildings do not require adherence to safety codes

Q40: How do feeder cables influence the overall energy efficiency of an electrical system?

A. They have no influence on energy efficiency
B. Properly sized and installed feeder cables reduce losses and improve system efficiency
C. Feeder cables decrease efficiency as they age

D. Only new cables improve efficiency

Q41: What are the consequences of failing to periodically inspect and maintain electrical feeder systems?

A. Systems become more efficient over time
B. Potential for system failures, safety hazards, and inefficient operation
C. Periodic inspections can damage the cables
D. Maintenance only affects the visual appeal of the system

Q42: What technologies are being implemented to monitor the condition of feeder cables in real time?

A. Basic visual inspection technologies
B. Advanced sensor networks that detect and report changes in electrical characteristics
C. No technologies are available for real-time monitoring
D. Use of manual checking methods only

Q43: How do changes in regulatory standards affect the installation and upgrade of feeder systems?

A. Standards have minimal impact on existing installations
B. New standards often require significant updates to systems to meet safety, efficiency, and environmental benchmarks
C. Regulatory changes only affect new installations
D. Changes in standards are only advisory, not mandatory

Q44: What are the best practices for decommissioning old feeder cables?

A. Leave them in place to maintain historical integrity
B. Safe removal, recycling of materials, and ensuring that new installations do not disturb the decommissioned cables
C. Reuse in different applications within the same facility
D. Burning them to ensure complete disposal

Q45: What type of protective device is typically used to prevent overloading in electrical feeder systems?

A. Grounding rod
B. Circuit breaker
C. Voltage regulator
D. Frequency modulator

46. Q46: In what situations is the use of aluminum conductors in feeder cables preferable to copper?

A. When higher conductivity is required
B. When lower cost and lighter weight are beneficial
C. When maximum flexibility is needed
D. When higher resistance to corrosion is necessary

Q47: How do environmental considerations affect the routing of feeder cables?

A. Cables must be routed to be aesthetically pleasing
B. Environmental factors like potential flooding, chemical exposure, or mechanical damage risk must be considered to ensure reliability
C. Routing is typically planned to create the shortest possible path, regardless of environment
D. Environmental considerations are secondary to cost considerations

Q48: What type of insulation is typically used for feeder cables in industrial applications?

A. Basic plastic insulation
B. High-grade insulation materials like cross-linked polyethylene (XLPE) for thermal and electrical resistance
C. Decorative insulation
D. Insulation is not required for industrial feeder cables

Q49: What are the typical maintenance routines for electrical feeders in a commercial building?

A. Daily cleaning and polishing
B. Regular inspections for wear and damage, and periodic testing for electrical performance
C. Maintenance routines are not necessary for modern systems
D. Only replaced when visible faults occur

Q50: How does the installation of feeder cables in public areas differ from private areas?

A. Public installations require more decorative cables
B. Public installations must adhere to stricter safety standards and are often required to be buried or placed in protective conduits
C. There is no difference in the installation process
D. Private installations require higher security measures

Q51: What is a feeder tap, and what are the regulations that must be considered when implementing one?

A. A decorative element in feeder systems
B. A connection to the main feeder cable that supplies power to a smaller branch circuit, subject to specific code requirements regarding its size and length
C. A type of switch used to control feeder lines
D. A monitoring device installed on feeders

52. Q52: How should feeder cables be rated in terms of current capacity relative to the expected load?

A. At exactly the same current as the expected load for efficiency
B. Lower than the expected load to save on costs
C. Higher than the expected load to allow for future expansion and avoid overheating
D. Randomly, as it does not affect performance

Q53: What role does the conductor's temperature rating play in the selection of feeder cables?

A. The temperature rating indicates the aesthetic value of the cable
B. It is crucial for determining the safe operating temperature range of the cable
C. Temperature rating is only important for outdoor cables
D. It determines the electrical resistance of the cable

Q54: How does the choice of conductor material impact the performance of feeder cables?

A. Only gold conductors should be used for optimal performance
B. Copper offers lower resistance than aluminum, which can affect efficiency and voltage drop
C. Conductor material has no impact on performance
D. All conductors perform the same under electrical load

55. Q55: What safety tests should be conducted on feeder cables before they are put into service?

A. Color testing to ensure they are visible
B. Insulation resistance tests and continuity checks to ensure there are no defects

C. Weight tests to ensure they can support physical loads
D. Aesthetic checks to ensure they match the environment

56. Q56: How does the length of a feeder cable affect its design and specification?

A. Longer cables require lighter colors to reduce heat absorption
B. Longer feeder cables might require larger cross-sectional areas to mitigate voltage drop and power loss
C. Length has no impact on the design of feeder cables
D. Shorter cables always require more insulation

57. Q57: What is the impact of feeder cable capacitance in high-frequency power transmission?

A. It increases the speed of electricity flow
B. Capacitance can lead to power losses and affect voltage regulation
C. Capacitance is only a concern in low-frequency applications
D. It makes the system more cost-effective

58. Q58: Why must feeder cables be resistant to UV light and chemicals?

A. UV and chemical resistance is only necessary for indoor cables
B. It ensures that the cables do not conduct electricity
C. Resistance to these elements prevents degradation and prolongs the life of the cable in harsh environments
D. It affects the color and design of the cables

59. Q59: How do environmental considerations affect the routing of feeder cables?

A. Cables must be routed to be aesthetically pleasing
B. Environmental factors like potential flooding, chemical exposure, or mechanical damage risk must be considered to ensure reliability
C. Routing is typically planned to create the shortest possible path, regardless of environment
D. Environmental considerations are secondary to cost considerations

60. Q60: What type of insulation is typically used for feeder cables in industrial applications?

A. Basic plastic insulation

B. High-grade insulation materials like cross-linked polyethylene (XLPE) for thermal and electrical resistance
C. Decorative insulation
D. Insulation is not required for industrial feeder cables

61. Q61: What are the typical maintenance routines for electrical feeders in a commercial building?

A. Daily cleaning and polishing
B. Regular inspections for wear and damage, and periodic testing for electrical performance
C. Maintenance routines are not necessary for modern systems
D. Only replaced when visible faults occur

62. Q62: How does the installation of feeder cables in public areas differ from private areas?

A. Public installations require more decorative cables
B. Public installations must adhere to stricter safety standards and are often required to be buried or placed in protective conduits
C. There is no difference in the installation process
D. Private installations require higher security measures

63. Q63: What is a feeder pillar and what role does it play in electrical distribution systems?

A. A decorative component of the electrical system
B. A central point for connecting different feeder cables
C. An advanced sensor for detecting current flow
D. A device that steps up the voltage in feeder cables

64. Q64: How are feeder lines protected from overcurrent and short circuits?

A. By using fuses and circuit breakers rated for the expected loads
B. By installing additional capacitors along the line
C. By decreasing the wire diameter
D. By manually monitoring the feeders continuously

Q65: What is voltage drop in feeder cables and why is it important to control it?

A. It enhances the efficiency of the electrical system
B. Voltage drop refers to the reduction in voltage as electricity

travels along the cable, affecting how devices function at the circuit's end
C. Voltage drop is a desired characteristic in residential installations
D. It refers to the physical sagging of cables over time

Q66: What are the consequences of improperly sized feeder cables?

A. Increased aesthetic appeal of the electrical system
B. Can lead to excessive voltage drop, overheating, and potential fire hazards
C. Makes the system more energy-efficient
D. Reduces the cost of materials

67. Q67: How should feeder cables be installed to ensure safety and compliance with electrical codes?

A. Loosely to allow for natural expansion and contraction
B. In a straight line regardless of the landscape
C. According to manufacturer's instructions and local code requirements, often in conduits or cable trays
D. With minimal securing to allow for easy re-routing

68. Q68: What is the role of a feeder breaker in an electrical distribution system?

A. To break the cable for easy installation
B. To manually turn circuits on and off
C. To interrupt power automatically in case of an overload or short-circuit
D. To provide a fixed connection point between different cable types

69. Q69: In what scenarios might a feeder need to be derated?

A. In low-temperature environments
B. When the feeder is exposed to high ambient temperatures or is bundled with other cables, potentially increasing heat buildup
C. When the feeder is under low load conditions
D. When the feeder is used in temporary installations

Q70: What considerations should be made for the expansion and contraction of feeder cables?

189

A. Cables should be painted to prevent expansion
B. Allowance for thermal expansion should be made in cable runs to prevent mechanical stress
C. Cables should be tightly clamped to prevent movement
D. Expansion is not a concern with modern cables

Q71: What technologies are being implemented to monitor the condition of feeder cables in real time?

A. Basic visual inspection technologies
B. Advanced sensor networks that detect and report changes in electrical characteristics
C. No technologies are available for real-time monitoring
D. Use of manual checking methods only

Q72: How do changes in regulatory standards affect the installation and upgrade of feeder systems?

A. Standards have minimal impact on existing installations
B. New standards often require significant updates to systems to meet safety, efficiency, and environmental benchmarks
C. Regulatory changes only affect new installations
D. Changes in standards are only advisory, not mandatory

Q73: What are the best practices for decommissioning old feeder cables?

A. Leave them in place to maintain historical integrity
B. Safe removal, recycling of materials, and ensuring that new installations do not disturb the decommissioned cables
C. Reuse in different applications within the same facility
D. Burning them to ensure complete disposal

Q74: What type of protective device is typically used to prevent overloading in electrical feeder systems?

A. Grounding rod
B. Circuit breaker
C. Voltage regulator
D. Frequency modulator

Q75: In what situations is the use of aluminum conductors in feeder cables preferable to copper?

A. When higher conductivity is required
B. When lower cost and lighter weight are beneficial

C. When maximum flexibility is needed
D. When higher resistance to corrosion is necessary

Q76: How does the presence of harmonics in the system affect feeder cable selection?

A. No effect, as harmonics are unrelated to feeders
B. Requires thicker insulation to prevent interference
C. Might need cables with higher current carrying capacity to handle increased heat
D. Shorter cable lengths are preferred to minimize effects

Q77: What is the significance of impedance in feeder cables?

A. Lower impedance helps in reducing energy costs
B. High impedance is necessary for better voltage regulation
C. Impedance impacts the voltage drop along the cable
D. Impedance affects the color and texture of the cable

Q78: Why is it crucial to ensure proper grounding in electrical feeder systems?

A. Grounding affects the operational speed of the circuit
B. Proper grounding prevents electrical shock and stabilizes voltage levels
C. Grounding is only necessary for aesthetic purposes
D. Grounding increases the energy efficiency of the system

Q79: What is the impact of long feeder runs on power quality, and how can it be mitigated?

A. Longer runs improve power quality by increasing voltage
B. Long runs may lead to voltage drops and increased impedance, mitigated by using cables with larger cross-sectional areas or installing repeaters
C. Long feeder runs have no impact on power quality
D. They decrease power quality by reducing current

Q80: How should feeder cables be routed to minimize interference with communication lines?

A. Parallel and close to communication lines to save space
B. At a safe distance or using shielded cables to prevent electromagnetic interference

C. Interference is not an issue with modern cables

D. Routing them randomly, as it does not matter

Q81: What are the specific code requirements that dictate the use and installation of feeder cables in residential versus commercial buildings?

A. Residential installations require less stringent codes
B. Both types of buildings follow the same codes for simplicity
C. Codes vary significantly, typically requiring more robust protection and redundancy in commercial settings
D. Commercial buildings do not require adherence to safety codes

Q82: How do feeder cables influence the overall energy efficiency of an electrical system?

A. They have no influence on energy efficiency
B. Properly sized and installed feeder cables reduce losses and improve system efficiency
C. Feeder cables decrease efficiency as they age
D. Only new cables improve efficiency

Q83: What are the consequences of failing to periodically inspect and maintain electrical feeder systems?

A. Systems become more efficient over time
B. Potential for system failures, safety hazards, and inefficient operation
C. Periodic inspections can damage the cables
D. Maintenance only affects the visual appeal of the system

Q84: What is the primary function of electrical feeders?

A. To regulate voltage in a circuit
B. To distribute power from one main source to various branch circuits
C. To convert AC to DC
D. To store electrical energy

Q85: What factor must be considered when sizing an electrical feeder for a new installation?

A. Color of the feeder cable
B. Total expected load of the circuit
C. Length of the cable only
D. Type of insulation material

Q86: What is a feeder pillar, and what role does it play in electrical distribution systems?

A. A decorative component of the electrical system
B. A central point for connecting different feeder cables
C. An advanced sensor for detecting current flow
D. A device that steps up the voltage in feeder cables

Q87: How should feeder cables be protected from mechanical damage, especially in industrial settings?

A. By covering them with decorative tape
B. By using robust conduit systems or armored cable where risk of damage is high
C. Protection is not necessary if cables are installed professionally
D. By painting them with protective coatings

Q88: What is the role of a feeder breaker in an electrical distribution system?

A. To break the cable for easy installation
B. To manually turn circuits on and off
C. To interrupt power automatically in case of an overload or short-circuit
D. To provide a fixed connection point between different cable types

Q89: What type of insulation is typically used for feeder cables in industrial applications?

A. Basic plastic insulation
B. High-grade insulation materials like cross-linked polyethylene (XLPE) for thermal and electrical resistance
C. Decorative insulation
D. Insulation is not required for industrial feeder cables

Q90: What are the typical maintenance routines for electrical feeders in a commercial building?

A. Daily cleaning and polishing
B. Regular inspections for wear and damage, and periodic testing for electrical performance
C. Maintenance routines are not necessary for modern systems
D. Only replaced when visible faults occur

Q91: How should feeder cables be installed to ensure safety and compliance with electrical codes?

A. Loosely to allow for natural expansion and contraction
B. In a straight line regardless of the landscape
C. According to manufacturer's instructions and local code requirements, often in conduits or cable trays
D. With minimal securing to allow for easy re-routing

Q92: How does the choice of conductor material impact the performance of feeder cables?

A. Only gold conductors should be used for optimal performance
B. Copper offers lower resistance than aluminum, which can affect efficiency and voltage drop
C. Conductor material has no impact on performance
D. All conductors perform the same under electrical load

Q93: What are the consequences of improperly sized feeder cables?

A. Increased aesthetic appeal of the electrical system
B. Can lead to excessive voltage drop, overheating, and potential fire hazards
C. Makes the system more energy-efficient
D. Reduces the cost of materials

Q94: How does the installation of feeder cables in public areas differ from private areas?

A. Public installations require more decorative cables
B. Public installations must adhere to stricter safety standards and are often required to be buried or placed in protective conduits
C. There is no difference in the installation process
D. Private installations require higher security measures

Q95: How are feeder lines protected from overcurrent and short circuits?

A. By using fuses and circuit breakers rated for the expected loads
B. By installing additional capacitors along the line
C. By decreasing the wire diameter
D. By manually monitoring the feeders continuously

Q96: What is voltage drop in feeder cables and why is it important to control it?

A. It enhances the efficiency of the electrical system
B. Voltage drop refers to the reduction in voltage as electricity travels along the cable, affecting how devices function at the circuit's end
C. Voltage drop is a desired characteristic in residential installations
D. It refers to the physical sagging of cables over time

Q97: What considerations should be made for the expansion and contraction of feeder cables?

A. Cables should be painted to prevent expansion
B. Allowance for thermal expansion should be made in cable runs to prevent mechanical stress
C. Cables should be tightly clamped to prevent movement
D. Expansion is not a concern with modern cables

Q98: How do environmental factors like humidity and salinity affect the choice of feeder cable materials?

A. They necessitate the use of gold-plated cables
B. Resistant materials such as tinned copper or moisture-resistant insulation are chosen to prevent corrosion and degradation
C. These factors enhance the conductivity of cables
D. Humidity and salinity are primarily concerns for wireless transmission

Q99: What is the impact of long feeder runs on power quality, and how can it be mitigated?

A. Longer runs improve power quality by increasing voltage
B. Long runs may lead to voltage drops and increased impedance, mitigated by using cables with larger cross-sectional areas or installing repeaters
C. Long feeder runs have no impact on power quality
D. They decrease power quality by reducing current

Q100: What technologies are being implemented to monitor the condition of feeder cables in real time?

A. Basic visual inspection technologies

- B. Advanced sensor networks that detect and report changes in electrical characteristics
- C. No technologies are available for real-time monitoring
- D. Use of manual checking methods only

Answers

Answer 1: B. To distribute power from one main source to various branch circuits.

Explanation: The primary function of electrical feeders is to distribute power from a central source to multiple branch circuits throughout a facility.

Answer 2: B. Total expected load of the circuit.

Explanation: When sizing an electrical feeder, the total expected load of the circuit must be considered to ensure the feeder can handle the demand.

Answer 3: C. Might need cables with higher current carrying capacity to handle increased heat.

Explanation: The presence of harmonics in the system can cause additional heat, requiring feeder cables with higher current carrying capacity.

Answer 4: C. Impedance impacts the voltage drop along the cable.

Explanation: Impedance affects how much voltage drops as electricity travels through the cable, which can impact the performance of electrical devices.

Answer 5: B. Proper grounding prevents electrical shock and stabilizes voltage levels.

Explanation: Grounding is crucial for safety, as it prevents electrical shocks and helps stabilize voltage levels in the system.

Answer 6: B. A central point for connecting different feeder cables.

Explanation: A feeder pillar serves as a central connection point for various feeder cables, facilitating distribution.

Answer 7: A. By using fuses and circuit breakers rated for the expected loads.

Explanation: Feeder lines are protected from overcurrent and short circuits using fuses and circuit breakers designed to handle the expected loads.

Answer 8: B. Voltage drop refers to the reduction in voltage as electricity travels along the cable, affecting how devices function at the circuit's end.

Explanation: Voltage drop is important to control because it impacts the performance of electrical devices by reducing the voltage available at the end of the cable.

Answer 9: B. Can lead to excessive voltage drop, overheating, and potential fire hazards.

Explanation: Improperly sized feeder cables can cause voltage drop issues, overheating, and create fire hazards.

Answer 10: C. According to manufacturer's instructions and local code requirements, often in conduits or cable trays.

Explanation: To ensure safety and compliance with electrical codes, feeder cables should be installed according to the manufacturer's instructions and local codes, often using conduits or cable trays.

Answer 11: C. To interrupt power automatically in case of an overload or short-circuit.

Explanation: A feeder breaker automatically interrupts power in the event of an overload or short-circuit to protect the system.

Answer 12: B. When the feeder is exposed to high ambient temperatures or is bundled with other cables, potentially increasing heat buildup.

Explanation: A feeder may need to be derated when exposed to high temperatures or bundled conditions that increase heat buildup.

Answer 13: B. Allowance for thermal expansion should be made in cable runs to prevent mechanical stress.

Explanation: Feeder cables should allow for thermal expansion to avoid mechanical stress and potential damage.

Answer 14: B. Copper offers lower resistance than aluminum, which can affect efficiency and voltage drop.

Explanation: Copper has lower resistance compared to aluminum, which can improve efficiency and reduce voltage drop in feeder cables.

Answer 15: B. Insulation resistance tests and continuity checks to ensure there are no defects.

Explanation: Safety tests such as insulation resistance tests and continuity checks are necessary to identify defects before putting feeder cables into service.

Answer 16: B. Longer feeder cables might require larger cross-sectional areas to mitigate voltage drop and power loss.

Explanation: The length of a feeder cable affects its design; longer cables often need a larger cross-sectional area to reduce voltage drop and power loss.

Answer 17: B. Capacitance can lead to power losses and affect voltage regulation.

Explanation: Feeder cable capacitance can result in power losses and affect voltage regulation, especially in high-frequency power transmission.

Answer 18: C. Resistance to these elements prevents degradation and prolongs the life of the cable in harsh environments.

Explanation: Feeder cables need to be resistant to UV light and chemicals to prevent degradation and extend their lifespan in harsh conditions.

Answer 19: B. Environmental factors like potential flooding, chemical exposure, or mechanical damage risk must be considered to ensure reliability.

Explanation: Routing feeder cables must take into account environmental factors to ensure the reliability and safety of the installation.

Answer 20: B. High-grade insulation materials like cross-linked polyethylene (XLPE) for thermal and electrical resistance.

Explanation: Industrial feeder cables typically use high-grade insulation materials like XLPE to provide necessary thermal and electrical resistance.

Answer 21: B. Regular inspections for wear and damage, and periodic testing for electrical performance.

Explanation: Maintenance routines for electrical feeders in commercial buildings include regular inspections and periodic testing to ensure proper performance.

Answer 22: B. Public installations must adhere to stricter safety standards and are often required to be buried or placed in protective conduits.

Explanation: Feeder cable installations in public areas require stricter safety standards and are often installed in protective conduits or buried.

Answer 23: B. A connection to the main feeder cable that supplies power to a smaller branch circuit, subject to specific code requirements regarding its size and length.

Explanation: A feeder tap connects to the main feeder cable to supply power to a smaller branch circuit and must adhere to specific code requirements.

Answer 24: C. Higher than the expected load to allow for future expansion and avoid overheating.

Explanation: Feeder cables should be rated higher than the expected load to accommodate future expansion and prevent overheating.

Answer 25: B. It is crucial for determining the safe operating temperature range of the cable.

Explanation: The temperature rating of a conductor is important as it determines the safe operating temperature range and ensures the cable performs reliably under expected conditions.

Answer 26: A. In situations requiring minimal visual impact and protection from environmental factors.

Explanation: Underground feeder (UF) cables are used in scenarios where aesthetics are important and where cables need protection from environmental factors.

Answer 27: C. Panels must accommodate the physical and electrical characteristics of feeder cables, including space for safe routing and connections.

Explanation: Feeder cables impact the design of electrical panels by requiring sufficient space and proper routing to accommodate their physical and electrical characteristics.

Answer 28: C. Electrical testing such as insulation resistance and continuity tests.

Explanation: After installation, feeder cables should be tested for insulation resistance and continuity to ensure they meet safety standards.

Answer 29: B. By using robust conduit systems or armored cable where risk of damage is high.

Explanation: Feeder cables in industrial settings should be protected using conduit systems or armored cable to safeguard against mechanical damage.

Answer 30: B. Long runs may lead to voltage drops and increased impedance, mitigated by using cables with larger cross-sectional areas or installing repeaters.

Explanation: Long feeder runs can cause voltage drops and increased impedance, which can be mitigated by using larger cables or installing repeaters.

Answer 31: B. Specific orientation and shielding strategies can help minimize EMI effects.

Explanation: In environments with high electromagnetic interference (EMI), proper orientation and shielding of feeder cables can help reduce EMI effects.

Answer 32: B. Use of flexible supports and expansion joints to allow for movement without damage.

Explanation: In seismic areas, feeder cables should be supported flexibly and use expansion joints to accommodate movement and prevent damage.

Answer 33: B. Resistant materials such as tinned copper or moisture-resistant insulation are chosen to prevent corrosion and degradation.

Explanation: In humid or saline environments, materials like tinned copper or moisture-resistant insulation are used to prevent corrosion and degradation.

Answer 34: B. Innovations like smart grid-compatible cables, higher temperature insulation materials, and improved conductor materials.

Explanation: Advancements in feeder cable technology include smart grid-compatible cables, higher temperature insulation, and improved conductor materials for better safety and efficiency.

Answer 35: C. Detailed labeling including source, destination, and power specifications to facilitate maintenance and troubleshooting.

Explanation: Feeder cables in large facilities should be detailed with labels indicating source, destination, and power specifications to aid maintenance and troubleshooting.

Answer 36: B. Regular maintenance checks, proper installation practices, and the use of suitable connectors and joints.

Explanation: Ensuring reliable feeder cable connections involves regular maintenance, proper installation, and using appropriate connectors and joints.

Answer 37: B. Managing variable power outputs and ensuring compatibility with grid standards.

Explanation: Integrating renewable energy sources with existing feeder infrastructure involves managing variable outputs and ensuring compatibility with grid standards.

Answer 38: B. At a safe distance or using shielded cables to prevent electromagnetic interference.

Explanation: To minimize interference with communication lines, feeder cables should be routed at a safe distance or use shielded cables.

Answer 39: C. Codes vary significantly, typically requiring more robust protection and redundancy in commercial settings.

Explanation: Regulatory codes for feeder cables vary between residential and commercial buildings, with commercial settings typically requiring more robust protection and redundancy.

Answer 40: B. Properly sized and installed feeder cables reduce losses and improve system efficiency.

Explanation: The efficiency of an electrical system is influenced by properly sized and installed feeder cables, which help reduce losses and improve overall efficiency.

Answer 41: B. Potential for system failures, safety hazards, and inefficient operation.

Explanation: Failing to periodically inspect and maintain electrical feeder systems can lead to system failures, safety hazards, and inefficient operation.

Answer 42: B. Advanced sensor networks that detect and report changes in electrical characteristics.

Explanation: Technologies for monitoring feeder cables in real time include advanced sensor networks that track and report changes in electrical characteristics.

Answer 43: B. New standards often require significant updates to systems to meet safety, efficiency, and environmental benchmarks.

Explanation: Changes in regulatory standards often necessitate updates to feeder systems to comply with new safety, efficiency, and environmental requirements.

Answer 44: B. Safe removal, recycling of materials, and ensuring that new installations do not disturb the decommissioned cables.

Explanation: Best practices for decommissioning old feeder cables include safe removal, recycling, and ensuring new installations do not interfere with old cables.

Answer 45: B. Circuit breaker.

Explanation: A circuit breaker is used to prevent overloading in electrical feeder systems by automatically interrupting power in the event of an overload.

Answer 46: B. When lower cost and lighter weight are beneficial.

Explanation: Aluminum conductors are preferred over copper when cost and lighter weight are important, despite having higher resistance.

Answer 47: B. Environmental factors like potential flooding, chemical exposure, or mechanical damage risk must be considered to ensure reliability.

Explanation: Routing of feeder cables must account for environmental factors to ensure their reliability and protection from potential risks.

Answer 48: B. High-grade insulation materials like cross-linked polyethylene (XLPE) for thermal and electrical resistance.

Explanation: Industrial applications typically use high-grade insulation materials such as XLPE for their superior thermal and electrical resistance.

Answer 49: B. Regular inspections for wear and damage, and periodic testing for electrical performance.

Explanation: Maintenance routines for electrical feeders in commercial buildings include regular inspections and periodic testing to ensure proper operation.

Answer 50: B. Public installations must adhere to stricter safety standards and are often required to be buried or placed in protective conduits.

Explanation: Feeder cable installations in public areas are subject to stricter safety standards and are often required to be buried or placed in protective conduits.

Answer 51: B. A connection to the main feeder cable that supplies power to a smaller branch circuit, subject to specific code requirements regarding its size and length.

Explanation: A feeder tap is a connection that branches off from the main feeder cable to supply power to a smaller circuit. It must adhere to specific code requirements for size and length.

Answer 52: C. Higher than the expected load to allow for future expansion and avoid overheating.

Explanation: Feeder cables should be rated higher than the expected load to accommodate future expansion and prevent overheating.

Answer 53: B. It is crucial for determining the safe operating temperature range of the cable.

Explanation: The conductor's temperature rating indicates the safe operating temperature range, which is important for preventing overheating and ensuring reliable operation.

Answer 54: B. Copper offers lower resistance than aluminum, which can affect efficiency and voltage drop.

Explanation: Copper conductors offer lower resistance compared to aluminum, which improves efficiency and reduces voltage drop.

Answer 55: B. Insulation resistance tests and continuity checks to ensure there are no defects.

Explanation: Before putting feeder cables into service, insulation resistance tests and continuity checks are essential to ensure there are no defects and that the cables are safe to use.

Answer 56: B. Longer feeder cables might require larger cross-sectional areas to mitigate voltage drop and power loss.

Explanation: Longer feeder cables often need larger cross-sectional areas to reduce voltage drop and minimize power loss.

Answer 57: B. Capacitance can lead to power losses and affect voltage regulation.

Explanation: In high-frequency power transmission, capacitance in feeder cables can cause power losses and impact voltage regulation.

Answer 58: C. Resistance to these elements prevents degradation and prolongs the life of the cable in harsh environments.

Explanation: Feeder cables must be resistant to UV light and chemicals to prevent degradation and extend their lifespan in harsh environments.

Answer 59: B. Environmental factors like potential flooding, chemical exposure, or mechanical damage risk must be considered to ensure reliability.

Explanation: Routing feeder cables requires consideration of environmental factors to ensure the cables' reliability and protection from potential risks.

Answer 60: B. High-grade insulation materials like cross-linked polyethylene (XLPE) for thermal and electrical resistance.

Explanation: Industrial applications typically use high-grade insulation materials such as XLPE for their excellent thermal and electrical resistance.

Answer 61: B. Regular inspections for wear and damage, and periodic testing for electrical performance.

Explanation: Maintenance routines for electrical feeders in commercial buildings include regular inspections and periodic testing to ensure proper performance and detect any issues.

Answer 62: B. Public installations must adhere to stricter safety standards and are often required to be buried or placed in protective conduits.

Explanation: Feeder cable installations in public areas are subject to stricter safety standards and are often required to be buried or placed in protective conduits to ensure safety.

Answer 63: B. A central point for connecting different feeder cables.

Explanation: A feeder pillar acts as a central point where different feeder cables are connected, facilitating electrical distribution.

Answer 64: A. By using fuses and circuit breakers rated for the expected loads.

Explanation: Feeder lines are protected from overcurrent and short circuits by using fuses and circuit breakers that are properly rated for the expected loads.

Answer 65: B. Voltage drop refers to the reduction in voltage as electricity travels along the cable, affecting how devices function at the circuit's end.

Explanation: Voltage drop is the decrease in voltage along the length of a cable, which can affect the performance of devices at the end of the circuit.

Answer 66: B. Can lead to excessive voltage drop, overheating, and potential fire hazards.

Explanation: Improperly sized feeder cables can result in excessive voltage drop, overheating, and pose fire hazards due to inadequate current-carrying capacity.

Answer 67: C. According to manufacturer's instructions and local code requirements, often in conduits or cable trays.

Explanation: Feeder cables should be installed according to manufacturer's instructions and local code requirements, typically using conduits or cable trays for protection and compliance.

Answer 68: C. To interrupt power automatically in case of an overload or short-circuit.

Explanation: A feeder breaker automatically interrupts power to protect the system in case of an overload or short-circuit.

Answer 69: B. When the feeder is exposed to high ambient temperatures or is bundled with other cables, potentially increasing heat buildup.

Explanation: Feeder cables may need to be derated when exposed to high ambient temperatures or when bundled with other cables to prevent excessive heat buildup.

Answer 70: B. Allowance for thermal expansion should be made in cable runs to prevent mechanical stress.

Explanation: Feeder cables should be installed with allowances for thermal expansion to avoid mechanical stress and damage.

Answer 71: B. Advanced sensor networks that detect and report changes in electrical characteristics.

Explanation: Technologies for monitoring feeder cables in real time include advanced sensor networks that track and report changes in electrical characteristics.

Answer 72: B. New standards often require significant updates to systems to meet safety, efficiency, and environmental benchmarks.

Explanation: Changes in regulatory standards often necessitate significant updates to systems to comply with new safety, efficiency, and environmental benchmarks.

Answer 73: B. Safe removal, recycling of materials, and ensuring that new installations do not disturb the decommissioned cables.

Explanation: Best practices for decommissioning old feeder cables involve safe removal, recycling of materials, and ensuring that new installations do not interfere with the old cables.

Answer 74: B. Circuit breaker.

Explanation: A circuit breaker is used to prevent overloading in electrical feeder systems by automatically interrupting power when necessary.

Answer 75: B. When lower cost and lighter weight are beneficial.

Explanation: Aluminum conductors are used in feeder cables when lower cost and lighter weight are advantageous, despite their higher resistance compared to copper.

Here are the answers for the new set of questions:

Answer 76: C. Might need cables with higher current carrying capacity to handle increased heat.

Explanation: Harmonics can increase the heat generated in feeder cables, so cables with higher current-carrying capacity may be required to handle this heat.

Answer 77: C. Impedance impacts the voltage drop along the cable.

205

Explanation: Impedance affects how much voltage drops along the length of a feeder cable, which can impact the performance of electrical devices at the end of the circuit.

Answer 78: B. Proper grounding prevents electrical shock and stabilizes voltage levels.

Explanation: Proper grounding is crucial for safety and stability, preventing electrical shock and maintaining stable voltage levels.

Answer 79: B. Long runs may lead to voltage drops and increased impedance, mitigated by using cables with larger cross-sectional areas or installing repeaters.

Explanation: Long feeder runs can cause voltage drops and increased impedance. This can be mitigated by using larger cables or installing repeaters to maintain power quality.

Answer 80: B. At a safe distance or using shielded cables to prevent electromagnetic interference.

Explanation: To minimize interference with communication lines, feeder cables should be routed at a safe distance or use shielded cables to reduce electromagnetic interference.

Answer 81: C. Codes vary significantly, typically requiring more robust protection and redundancy in commercial settings.

Explanation: Residential and commercial buildings follow different code requirements, with commercial settings usually requiring more robust protection and redundancy.

Answer 82: B. Properly sized and installed feeder cables reduce losses and improve system efficiency.

Explanation: Correctly sized and installed feeder cables help to minimize energy losses and improve overall system efficiency.

Answer 83: B. Potential for system failures, safety hazards, and inefficient operation.

Explanation: Failing to periodically inspect and maintain electrical feeder systems can lead to system failures, safety hazards, and inefficient operation.

Answer 84: B. To distribute power from one main source to various branch circuits.

Explanation: The primary function of electrical feeders is to distribute power from a main source to multiple branch circuits.

Answer 85: B. Total expected load of the circuit.

Explanation: When sizing an electrical feeder, the total expected load of the circuit is a crucial factor to ensure the cable can handle the load safely.

Answer 86: B. A central point for connecting different feeder cables.

Explanation: A feeder pillar acts as a central connection point for various feeder cables in an electrical distribution system.

Answer 87: B. By using robust conduit systems or armored cable where risk of damage is high.

Explanation: To protect feeder cables from mechanical damage in industrial settings, robust conduit systems or armored cables should be used.

Answer 88: C. To interrupt power automatically in case of an overload or short-circuit.

Explanation: A feeder breaker automatically interrupts power to protect the system during an overload or short-circuit situation.

Answer 89: B. High-grade insulation materials like cross-linked polyethylene (XLPE) for thermal and electrical resistance.

Explanation: Industrial applications typically use high-grade insulation materials such as XLPE for their excellent thermal and electrical resistance properties.

Answer 90: B. Regular inspections for wear and damage, and periodic testing for electrical performance.

Explanation: Typical maintenance routines for electrical feeders in commercial buildings include regular inspections and periodic testing to ensure proper operation and identify any issues.

Answer 91: C. According to manufacturer's instructions and local code requirements, often in conduits or cable trays.

Explanation: Feeder cables should be installed according to manufacturer's instructions and local code requirements, commonly using conduits or cable trays for protection.

Answer 92: B. Copper offers lower resistance than aluminum, which can affect efficiency and voltage drop.

Explanation: Copper conductors generally offer lower resistance compared to aluminum, which can impact efficiency and reduce voltage drop.

Answer 93: B. Can lead to excessive voltage drop, overheating, and potential fire hazards.

Explanation: Improperly sized feeder cables can result in excessive voltage drop, overheating, and potential fire hazards due to inadequate current-carrying capacity.

Answer 94: B. Public installations must adhere to stricter safety standards and are often required to be buried or placed in protective conduits.

Explanation: Feeder cables installed in public areas must meet stricter safety standards and are typically required to be buried or placed in protective conduits.

Answer 95: A. By using fuses and circuit breakers rated for the expected loads.

Explanation: Feeder lines are protected from overcurrent and short circuits by employing fuses and circuit breakers that are rated according to the expected loads.

Answer 96: B. Voltage drop refers to the reduction in voltage as electricity travels along the cable, affecting how devices function at the circuit's end.

Explanation: Voltage drop is the reduction in voltage along a cable, which affects the performance of devices at the end of the circuit.

Answer 97: B. Allowance for thermal expansion should be made in cable runs to prevent mechanical stress.

Explanation: Feeder cables should be installed with allowances for thermal expansion to avoid mechanical stress and damage.

Answer 98: B. Resistant materials such as tinned copper or moisture-resistant insulation are chosen to prevent corrosion and degradation.

Explanation: Environmental factors like humidity and salinity require the use of resistant materials such as tinned copper or moisture-resistant insulation to prevent corrosion and degradation.

Answer 99: B. Long runs may lead to voltage drops and increased impedance, mitigated by using cables with larger cross-sectional areas or installing repeaters.

Explanation: Long feeder runs can result in voltage drops and increased impedance. This can be mitigated by using larger cables or installing repeaters to maintain power quality.

Answer 100: B. Advanced sensor networks that detect and report changes in electrical characteristics.

Explanation: Technologies for monitoring feeder cables in real time include advanced sensor networks that detect and report changes in electrical characteristics.